高等职业教育农业农村部"十三五"规划教材
"十三五"江苏省高等学校重点教材（2019-2-110）

动物繁殖

张响英　杨晓志　主编

Animal Reproduction

中国农业出版社
北　京

《动物繁殖》

内容简介

本教材系统介绍了动物的繁殖机理和实用繁殖技术,主要内容包括:采精、精液处理、发情鉴定及输精、胚胎移植、妊娠诊断、接产与助产、繁殖力评价,配有丰富的数字化教学资源,直观、真实、形象地呈现教学内容,激发学生的学习兴趣,利于加深学生对动物繁殖技术的理解。全文采用英汉对照,内容精炼,文字通俗,图文并茂。

Animal Reproduction

INTRODUCTION

This textbook systematically introduces the reproductive mechanism and practical reproductive technology of animals. The main contents include: semen collection, semen treatment, estrus identification and insemination, embryo transfer, pregnancy diagnosis, delivery and midwifery, evaluation of fecundity. And it is rich in digital resources, which are intuitive, real and vivid to present the teaching content, stimulate interest in learning, and deepen understanding of animal reproduction technology. The full textbook is in Chinese and English, the contents are refined, the languages are easy to understand, the pictures and texts are excellent.

《动物繁殖》

编 审 人 员

主　　编　张响英　杨晓志
副 主 编　张　蕾　王　靓
编　　者　（以姓氏笔画为序）
　　　　　王　靓　王利刚　杨晓志　何建文　张　蕾
　　　　　张响英　张海兰　张海波　陆艳凤　陈静波
　　　　　唐　鹏　甄　霆
行业指导　杜婷婷　岳丽娟
主　　审　李惠侠

Animal Reprduction

EDITORS

Chief Editor：Zhang Xiangying and Yang Xiaozhi

Associate Editor：Zhang Lei and Wang Liang

Editors：Wang Liang，Wang Ligang，Yang Xiaozhi，He Jianwen，
　　　　　Zhang Lei，Zhang Xiangying，Zhang Hailan，Zhang Haibo，
　　　　　Lu Yanfeng，Chen Jingbo，Tang Peng，Zhen Ting

Industry Instructor：Du Tingting and Yue Lijuan

Chief Reviewer：Li Huixia

PREFACE 前言

在经济全球化、文化多样化和社会信息化背景下，畜牧业的发展更加重视现代畜牧技术的交流和跨文化技术技能人才的培养与流动。中国"一带一路"倡议的实施，更是为中国和"一带一路"沿线国家和地区间畜牧业合作与发展、交流与互通创造了前所未有的机遇。现代畜牧业职业教育的国际化、专业技术技能人才的国际能力培养以及来华留学人员专业技能的培养，成为助推畜牧业技术、人才、装备等实现国际交流与合作的主要动力，人才培养依然是主体。专业技术能力、跨文化适应能力、国际交流能力和国际就业能力是人才培养必须关注的焦点，为适应这种需求而编写的汉英双语教材应运而生，成为推进畜牧业教育国际化的有益尝试和创新。

"动物繁殖"是畜牧兽医专业的一门技术课程，解密了动物新个体诞生的奥妙，诠释了人类为充分发挥动物的生殖潜能，在理念和技术手段上的智慧与发展历程。编写团队紧跟产业发展趋势

With the development of animal husbandry under the background of economic globalization, cultural diversity and social informatization, we have paid more attention to the exchange of modern animal husbandry technology and the cultivation of cross-cultural technical talents. Especially with the implementation of China's "The Belt and Road" initiative, it has created unprecedented opportunities for cooperation and exchanges in animal husbandry between China and the country "along the way". The internationalization of modern animal husbandry vocational education, the cultivation of international ability of technical talents and the cultivation of professional skills of overseas students have become the main driving force for promoting international exchanges and cooperation in technology, talents, and equipment. And talent training is still the key member of it. In the process of talent training, we always focus on the cultivation of technical ability, cross-cultural adaptability, international communication ability and inter-national employability. The bilingual textbook in Chinese and English is written to meet this demand. The compilation of the teaching material is also a beneficial attempt and innovation to promote the internationalization of animal husbandry education.

"Animal reproduction" is a technical course of animal husbandry and veterinary specialty, which deciphers the mystery of the birth of new animals, explains how human beings can make full use of the reproductive potential of animals with wisdom and technology. Our team closely follows the industrial development trend

和行业人才需求，严格按照岗位规范，坚持工学结合、知行合一，构建"模块化"知识体系。以典型的工作任务为驱动，以工作流程为主线确定编写内容，共分为采精、精液的处理、发情鉴定与输精、胚胎移植、妊娠诊断、接产与助产、繁殖力评价7个项目，可带领学生由浅入深了解动物繁殖的奥妙，由易及难逐步掌握动物繁殖所需的知识和技能。

教材编写过程中，紧扣国内外现代农牧业转型升级需求，考虑不同国家留学生的文化及宗教背景差异，适当选取相关的知识和技能，并吸收行业、企业一线的技术骨干、能工巧匠深度参与，引入行业发展的新工艺、新方法、新流程、新规范、新标准，以"必需""够用"为度，主要培养学生的核心岗位能力。对于工作步骤、操作流程尽可能采用图解、图示法，将图片进行有序编排，尽可能做到图文并茂。

本教材由张响英、杨晓志担任主编，负责全书的统稿和审核工作。项目一由陆艳凤、陈静波编写，项目二、项目七由张海波、王靓编写，项目三由张蕾、王利刚编写，项目四由张响英、张海兰编写，项目五由杨晓志、唐鹏编写，项目六由甄霆、何建文编写。本书在编写过程中，得到一线技术骨干杜婷婷、岳丽娟和外

and talent demand, strictly follows the post specification, adheres to the combination of work and learning, knowledge and practice, and builds a "modular" knowledge system. The content puts the typical tasks to the centre and makes the workflow a main line. It can be divided into 7 items: semen collection, semen treatment, estrus identification and insemination, embryo transfer, pregnancy diagnosis, delivery and midwifery, evaluation of fecundity. It can lead students to understand the mystery of animal reproduction, and master the knowledge and skills easily.

In the process of compiling the textbook, we closely follow the requirements for the transformation and upgrading of modern agriculture and animal husbandry, take into account the differences in cultural and religious backgrounds of overseas students, appropriately select relevant knowledge and skills, invite the technical key mernkers from the industry and enterprises to participate in the compilation, so as to introduce the new technology, new method, new process, new specification and new technology of the industry standard into the textbook. The selection of content is based on "necessary" and "sufficient". As far as possible, the operation procedures is decomposed and displayed through pictures, and pictures is arranged in order to facilitate students' learning.

The cheif editors of this textbook are Zhang Xiangying and YangXiaozhi, who are responsible for the compilation and check of the whole book. Project 1 was compiled by Lu Yanfeng and Chen Jingbo, project 2 and 7 by Zhang Haibo and Wang Liang, project 3 by Zhang Lei and Wang Ligang, project 4 by Zhang Xiangying and Zhang Hailan, project 5 by Yang Xiaozhi and Tang Peng, and project 6 by Zhen Ting and He Jianwen. In the process of compiling, we got a lot of guidance from technical experts from forefront of pro-

籍教师 Rosie Barnes 的大力帮助。南京农业大学李惠侠教授对本教材进行了审阅并提出了修改意见，在此一并表示衷心的感谢。

由于编者水平有限，如有疏漏与不当之处，恳请专家和广大读者批评指正。

编　者

2020 年 4 月

duction such as Du Tingting, Yue Lijuan and foreign teacher Rosie Barnes. Professor Li Huixia from Nanjing Agricultural University reviewed it. We express thanks to all of them.

Due to the limited level of editors, there must be mistakes although we tried our best to avoid them. Please point them out so that they can be corrected.

Editors

April 2020

CONTENTS 目 录

前言/ PREFACE

项目一 采精 ··· 1
Project I　Semen Collection ··· 1

 任务　采精 ·· 9
 Task　Semen Collection ··· 9

项目二 精液的处理 ·· 19
Project II　Semen Treatment ··· 19

 任务1　精液的品质评定 ··· 35
 Task 1　Evaluation of Semen ··· 35

 任务2　精液的稀释 ··· 44
 Task 2　Dilution of Semen ··· 44

 任务3　冷冻精液的制作 ··· 46
 Task 3　Production of Frozen Semen ··· 46

 任务4　精液的保存与运输 ·· 49
 Task 4　Preservation and Transportation of Semen ······································· 49

项目三 发情鉴定与输精 ··· 54
Project III　Estrus Identification and Insemination ······································ 54

 任务1　母牛的发情鉴定与输精 ··· 73
 Task 1　Estrus Identification and Insemination of Cows ································· 73

 任务2　母羊的发情鉴定与输精 ··· 80
 Task 2　Estrus Identification and Insemination of Ewes ································· 80

 任务3　母猪的发情鉴定与输精 ··· 82
 Task 3　Estrus Identification and Insemination of Sows ································· 82

 任务4　鸡的人工输精 ·· 87
 Task 4　Insemination of Hens ··· 87

项目四　胚胎移植 ·· 90
Project Ⅳ　Embryo Transfer ·· 90

任务 1　羊的胚胎移植 ·· 100
Task 1　Embryo Transfer in Sheep ··· 100
任务 2　牛的非手术法胚胎移植 ··· 103
Task 2　Non-surgical Embryo Transfer in Cattle ··· 103

项目五　妊娠诊断 ··· 120
Project Ⅴ　Pregnancy Diagnosis ·· 120

任务 1　牛的妊娠诊断 ··· 135
Task 1　Pregnancy Diagnosis of Cattle ··· 135
任务 2　羊的妊娠诊断 ··· 145
Task 2　Pregnancy Diagnosis of Sheep ··· 145
任务 3　猪的妊娠诊断 ··· 148
Task 3　Pregnancy Diagnosis of Pigs ·· 148

项目六　接产与助产 ·· 152
Project Ⅵ　Delivery and Midwifery ·· 152

任务 1　牛的接产与助产 ·· 161
Task 1　Delivery and Midwifery of Cattle ·· 161
任务 2　羊的接产与助产 ·· 168
Task 2　Delivery and Midwifery of Sheep ·· 168
任务 3　猪的接产与助产 ·· 173
Task 3　Delivery and Midwifery of Pigs ·· 173

项目七　繁殖力评价 ·· 177
Project Ⅶ　Evaluation of Fecundity ·· 177

任务　繁殖力评价 ··· 181
Task　Evaluation of Fecundity ·· 181

参考文献/References ·· 185

项目一 采 精
Project Ⅰ　Semen Collection

项目导学

采精是人工授精技术的首要环节，认真做好采精准备，掌握正确的采精步骤，合理安排采精频率，才能获得量多质优的精液，从而提高种公畜（禽）的利用效率。采精的方法很多，如假阴道法（牛、羊和兔等）、手握法（猪）、按摩法（禽类）和电刺激法等。

Project Guidance

Semen collection is the primary component of artificial insemination. In order to improve the utilization efficiency of male animals (poultry) and obtain high-quality semen, we must seriously prepare, correctly grasp the steps and reasonably arrange the frequency of semen collection. There are many methods for semen collection, such as the artificial vagina method (cattle, sheep, rabbits, etc.), the hand holding method (pigs), the massage method (poultry) and the electrical stimulation method.

学习目标

>>> 知识目标

• 了解雄性动物生殖器官的组成及组织结构特点。
• 掌握雄性动物生殖器官的主要生理功能。
• 理解精子的发生过程。

>>> 技能目标

• 熟练掌握假阴道的安装与调试。
• 规范采集公牛、公羊的精液。
• 熟练采集公猪的精液。
• 顺利完成公禽的采精。

Learning Objectives

>>> **Knowledge Objectives**

• To understand the composition and structural characteristics of male reproductive organs.
• To master the main physiological functions of male reproductive organs.
• To understand the process of spermatogenesis.

>>> **Skill Objectives**

• To master the installation and adjustment of artificial vagina.
• To collect sperm of bulls and rams using standard methods.
• To collect sperm of boars skillfully.
• To collect sperm of poultry successfully.

动物繁殖
Animal Reproduction

相关知识 | Relevant Knowledge

一、雄性动物生殖器官的结构与功能

雄性动物的生殖器官由睾丸、附睾、阴囊、输精管、副性腺、尿生殖道、阴茎和包皮组成。各种雄性动物的生殖器官见图1-1。

1　Structure and functions of male reproductive organs

The male reproductive organs consist of the testis, epididymis, scrotum, duct deferens, accessory sexual glands, urethra, penis and prepuce (Figure 1-1).

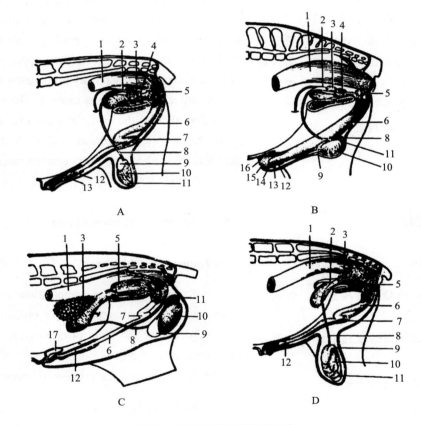

图1-1　雄性动物生殖器官示意

Figure 1-1　Reproductive organs of male animals

A. 牛　B. 马　C. 猪　D. 羊

A. bull　B. stallions　C. boar　D. ram

1. 直肠　2. 输精管壶腹　3. 精囊腺　4. 前列腺　5. 尿道球腺　6. 阴茎　7. S状弯曲　8. 输精管　9. 附睾头　10. 睾丸　11. 附睾尾　12. 阴茎游离端　13. 内包皮鞘　14. 外包皮鞘　15. 龟头　16. 尿道突起　17. 包皮憩室

1. rectum　2. ampulla of duct deferens　3. seminal vesicle gland　4. prostate gland　5. bulbourethral gland　6. penis　7. S-shaped bending　8. duct deferens　9. epididymal head　10. testis　11. epididymal tail　12. penis free end　13. internal prepuce sheath　14. outside prepuce sheath　15. glans penis　16. urethra protrusions　17. prepuce diverticulum

（一）睾丸

1. 形态和位置

睾丸为雄性动物的性腺，成对存在，呈卵圆形或长卵圆形。不同种属的雄性动物睾丸的大小和重量有较大差别，猪的睾丸相对较大，为900～1 000g，占体重的0.34%～0.38%；牛的相对较小，为550～650g，占体重的0.08%～0.09%。禽类睾丸大小和色泽因品种、年龄、生殖季节而有很大变化。雏鸡的睾丸只有米粒大小，呈淡黄色或黄色；成年公鸡在春季性机能特别旺盛，精子大量形成，睾丸颜色变白，体积变大；当性机能减退时体积变小。

睾丸原位于腹腔内、肾两侧，在胎儿期，由腹腔下降入阴囊。成年公畜如果一侧或两侧睾丸并未下降进入阴囊，称为隐睾，其内分泌机能不受损害，但精子生成会出现异常。公禽的睾丸位于体腔内，形似蚕豆。

2. 组织构造

睾丸表面被覆浆膜，其内为致密结缔组织构成的白膜。白膜由睾丸头端形成一条结缔组织索伸向睾丸实质，构成睾丸纵隔，将睾丸实质分成许多小叶。睾丸小叶由精曲小管盘曲构成，在各小叶顶端汇合成精直小管，穿入纵隔结缔组织内形成睾丸网（图1-2）。

1.1 Testis

1.1.1 Morphology and location

The testis is the gonad of male animals. It always exists in pairs and the shape is oval or oblong oval. There are many differences in the size and weight of testis among different male animals. The testis of boars are relatively large, which is 900-1 000 g, accounting for 0.34%-0.38% of the body weight. The testis of bulls are relatively small, which is 550-650 g, accounting for 0.08%-0.09% of the body weight. The size and color of poultry testis vary greatly among species, age and reproductive season. The chicken's testis is only a grain size, with a light yellow or yellow colour. In spring, the sexual function of adult cocks is vigorous, with a large number of sperm formed, and the testis become white and large. When the sexual function is going down, the testis becomes smaller.

The testis are originally located in the abdominal cavity, on both sides of the kidneys. During the fetal period, it descends from the abdominal cavity into scrotum. It is called cryptorchidism if one or both testis of adult male animals do not descend into the scrotum. The endocrine function of testis is not affected, but spermatogenesis will be abnormal. The testis of poultry is located in the abdominal cavity and looks like broad beans.

1.1.2 Tissue structure

The surface of testis is covered with serosa, which is a white membrane formed by dense connective tissue. The tunica albuginea forms a connective tissue from the head of the testis to the testicular parenchyma, which constructs the testicu-lar mediastinum and divides the testicular parenchyma into many lobules. The testicular lobules are composed of seminiferous tubules, which converge at the top of each lobule and penetrate into the connective tissue of the mediastinum to form the rete testis (Figure 1-2).

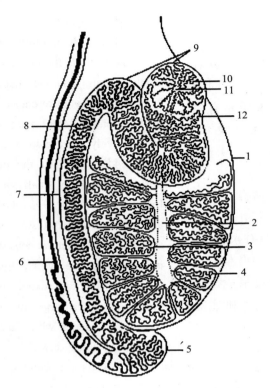

图 1-2 睾丸及附睾的组织构造

Figure 1-2　Tissue structure of testis and epididymis

1. 睾丸　2. 纵隔　3. 精曲小管　4. 小叶　5. 附睾尾　6. 输精管　7. 附睾体　8. 附睾管　9. 附睾头
10. 精直小管　11. 输出管　12. 睾丸网

1. testis　2. mediastinum　3. seminiferous tubules　4. foliole　5. epididymal tail　6. duct deferens
7. epididymal body　8. epididymal duct　9. epididymal head　10. tubulus rectus　11. duct efferent　12. rete testis

3. 机能

（1）产生精子。精曲小管生殖上皮的生精细胞（精原细胞）经增殖、分裂，最后形成精子。公牛每克睾丸组织平均每天可产生精子 1 300 万～1 900 万个，公猪 2 400 万～3 100 万个，公羊 2 400 万～2 700 万个，公马 1 930 万～2 200 万个。

（2）分泌雄激素。睾丸的间质细胞能分泌雄激素，可激发雄性动物的性欲及性行为，刺激第二性征，促进生殖器官的发育，维持精子发生及附睾内精子的存活。

1.1.3　Function

（1）Producing sperm. The spermatocytes in the germinal epithelium of seminiferous tubule proliferate, divide, and finally form sperm. Each gram of testicular tissue in a bull can produce 13-19 million sperm per day, 24-31 million in boars, 24-27 million in rams, and 19.3-22 million in stallions.

（2）Androgens secretion. The interstitial cells of the testis can secrete androgens, activating sexual desire and behavior in the male animals, stimulati-ng secondary sexual characteristics, promoting the development of genital organs, maintaining spermatogenesis and sperm survival in epididymis.

（二）附睾

1. 形态结构

附睾位于睾丸的附着缘，由头、体、尾三部分组成。附睾头膨大，由10~30条睾丸输出管盘曲组成。这些输出管汇集成一条较粗且弯曲的附睾管，构成附睾体。在睾丸的远端，附睾体延续并转为附睾尾，最后逐渐过渡为输精管。

2. 机能

睾丸是精子贮存和成熟的场所。从睾丸生成的精子，刚进入附睾头时颈部常有原生质滴，活动微弱，没有受精能力或受精能力很低。在精子通过附睾的过程中，原生质滴向尾部末端移行并脱落，精子逐渐成熟。由于附睾内的弱酸性（pH为6.2~6.8）、高渗透压、较低温度及厌氧的内环境，精子处于休眠状态。同时附睾管分泌物可提供给精子营养，因此精子在附睾内可贮存较长时间。

（三）输精管

输精管由附睾管延续而来，它与通往睾丸的神经、血管、淋巴管、睾丸提肌共同组成精索。输精管壁具有发达的平滑肌纤维，射精时凭借其强有力的收缩作用将精子排出。

（四）副性腺

1. 形态结构

副性腺包括精囊腺、前列腺和尿道球腺（图1-3）。公禽没有副性腺。精囊腺成对存在，位于输精管末端的外侧。前列腺位于精囊腺的后方，由体部和扩散部两部分组成。尿道球腺成对存在，位于尿生殖道骨盆部后端。

1.2 Epididymis

1.2.1 Morphology and structure

The epididymis is located at the attachment edge of the testis and consists of three parts: head, body and tail. The epididymal head is enlarged and consists of 10-30 testicular efferent tubes coiled. These tubes condense into a thick and curved epididymal tube, forming the epididymal body. At the far end of the testis, the epididymal body continues and turns into the epididymal tail, and eventually transits to duct deferens.

1.2.2 Function

Testis is the place where sperm are stored and matured. Sperms produced from testis often have protoplasmic droplets in the neck when they enter the epididymal head, with weak motility and have no or low fertility. The protoplasmic droplets fall off when the sperms pass through the epididymis, and sperms mature gradually. Because of the weak acidity (pH 6.2-6.8), high osmotic pressure, low temperature and anaerobic internal environment in the epididymis, the sperms are dormant. At the same time, epididymal duct secretions can provide nutrition for sperm, so sperm can be stored in epididymis for a long time.

1.3 Duct deferens

The duct deferens is a continuation of the epididymal duct, which together with the nerves, blood vessels, lymphatic vessels and the testicular muscles form the spermatic cord. The wall of the duct deferens has well-developed smooth muscle fibers, which expel sperm by virtue of its powerful contraction during ejaculation.

1.4 Accessory sexual glands

1.4.1 Morphology and structure

Accessory sexual glands include seminal vesicle gland, prostate gland and bulbourethral gland (Figure 1-3). There is no accessory sexual glands in the poultry. A pair of seminal vesicles glands are located at the outer side of the end of duct deferens. The prostate gland is located behind seminal vesicle gland and consists with body and diffused part. A pair of bulboure-

2. 机能

公畜在射精时，副性腺分泌物与输精管分泌物混合形成精清，有稀释精子、冲洗尿生殖道、活化精子、为精子提供营养等作用。

1.4.2 Function

When ejaculating, the secretion of accessory sexual glands and duct deferens is mixed to form seminal plasma, which has the functions of diluting sperm, rinsing urogenital tract, activating sperm, providing nutrition for sperm.

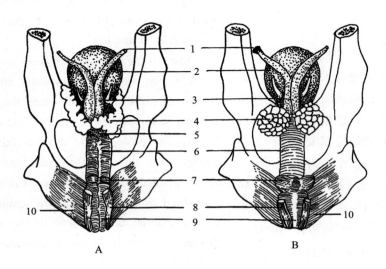

图 1-3 雄性动物的副性腺

Figure 1-3　Accessory sexual glands of male animals

A. 公牛　B. 公羊

A. bull　B. ram

1. 输精管　2. 膀胱　3. 壶腹　4. 精囊腺　5. 前列腺　6. 尿生殖道骨盆部
7. 尿道球腺　8. 坐骨海绵体肌　9. 阴茎缩肌　10. 球海绵体肌

1. duct deferens　2. bladder　3. ampullae　4. seminal vesicle gland　5. prostate gland
6. urogenital tract pelvis　7. bulbourethral gland　8. sciatic sponge muscle　9. penis muscle　10. bulbocavernosus muscle

（五）尿生殖道

尿生殖道为尿液和精液的共同通道，起源于膀胱，终于龟头，由骨盆部和阴茎部组成（图 1-4）。精阜主要由海绵体组织构成，在射精时可以膨大，关闭膀胱颈，阻止精液流入膀胱，同时阻止尿液混入精液。

1.5　Canalis urogenitalis

The canalis urogenitalis is a common channel for urine and semen. It originates from the bladder and ends in the glans, consists of the pelvis and penis (Figure 1-4). The verumontanum is mainly composed by the sponge tissue, which can be enlarged during ejaculation, close the bladder neck, prevent the semen from flowing into the bladder, and prevent urine from mixing into semen.

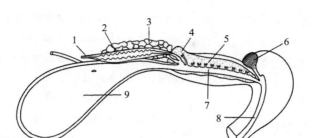

图 1-4　尿生殖道结构示意

Figure 1-4　Structure of urogenital tract

1. 输精管　2. 输精管壶腹　3. 精囊腺　4. 前列腺体部　5. 前列腺扩散部
6. 尿道球腺　7. 尿生殖道　8. 阴茎部　9. 膀胱

1. duct deferens　2. ampullae of duct deferens　3. seminal vesicle gland　4. body of prostate gland
5. diffuse part of prostate gland　6. bulbourethral gland　7. urogenital tract　8. penis　9. bladder

（六）阴茎和包皮

阴茎为雄性动物的交配器官。阴茎由阴茎海绵体和尿生殖道阴茎部组成，分为阴茎根、阴茎体和阴茎头。阴茎平时柔软，隐藏于包皮内。交配时勃起，伸长并变得粗硬。

二、精子的发生

精子的发生是指精子在睾丸内产生的全过程，主要历经以下三个阶段（图 1-5）。

1. 精原细胞的分裂增殖及精母细胞的形成

这一阶段历时 15～17d。其主要特点为：精原细胞的首次分裂同时产生一个活动的精原细胞和另一个暂时休眠的干细胞，所有的细胞分裂都是有丝分裂。理论上，一个 A_1 型精原细胞经几次分裂可生成 16 个初级精母细胞。

2. 精母细胞发育

初级精母细胞形成后，便进入静止期，期间细胞进行 DNA 复制，此后便进入成熟分裂，初级精母细胞分裂为 2

1.6　Penis and praeputium

The penis is the male reproductive organ of copulation. It consists of the corpus cavernosum and urogenital tract penis, which is divided into root, body and head of the penis. The penis is usually soft and protected in the foreskin. The penis elongates and becomes hard and fully erect in copulation.

2　Spermatogenesis

Spermatogenesis is the whole process of sperm production in testis, which mainly goes through three stages(Figure 1-5).

2.1　The proliferation of spermatogonia and the formation of spermatocytes

This stage lasts for 15-17 days and the main features are as follows. The first division of spermatogonia produces an active spermatocyte and another temporarily dormant stem cell. All cell divisions are mitosis. Theoretically, an A_1 spermatogonium can produce 16 primary spermatocytes after several divisions.

2.2　Development of spermatocytes

After the formation of primary spermatocytes, they enter a static phase which the cells undergo DNA replication and then enter mature division. A primary spermatocyte divides into two secondary spermatocytes,

个次级精母细胞，染色体数目减半。此阶段历时15～16d。次级精母细胞存在的时间很短，在一天之内分裂成2个精子细胞。

3. 精子的形成

精子细胞不再分裂而是经过复杂的形态结构变化形成精子。最初的精子细胞为圆形，以后逐渐变长，发生形态上的急剧变化，最后成为蝌蚪样的精子。此阶段历时10～15d。

在精子形成的过程中，由一个A_1型活动的精原细胞经过增殖、生长、成熟分裂及变形等阶段，最后形成精子这一过程所需的时间，称为精子发生周期。不同动物的精子发生周期不同，猪为44～45d,牛为54～60d,绵羊为49～50d,山羊为60d左右，马为49～50d。

and the number of chromosomes is halved. This stage lasts for 15-16 days. The presence of secondary spermatocytes is very short, and it splits into two sperm cells within one day.

2.3 Formation of sperm

Sperm cells no longer divide but evolve into sperm through complex morphological and structural changes. Initially, the sperm cells are round, and then gradually become longer, and the morphology changes dramatically. Finally, they become "tadpole" sperm. This stage lasts for 10-15days.

The time required for spermatogenesis is called spermatogenesis cycle, which goes through the proliferation, growth, mature division and deformation of an A_1 active spermatocytes. The spermatogenesis cycle is 44-45 days for boar, 49-50 days for ram, about 60 days for goat, 49-50 days for stallion.

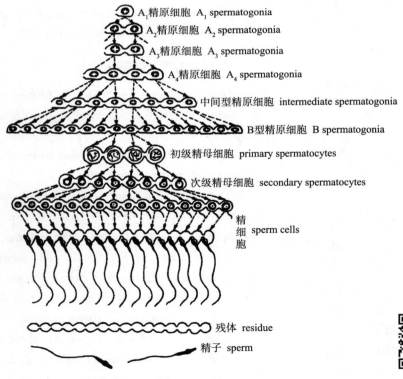

图1-5 精子发生过程示意

Figure 1-5 Spermatogenesis

精子的发生

Spermatogenesis

项目一 采 精

Project I Semen Collection

任 务 采 精
Task Semen Collection

任务描述

科学合理的采精技术是鲜精产量增加及质量提高的重要环节，有利于保持种公畜（禽）神经兴奋的持续性和连贯性。生产中，要根据种公畜（禽）的年龄、季节、体况等因素，严格规范采精操作，合理安排采精频率。

Task Description

Scientific and reasonable technology of semen collection is an important link in increasing the yield and quality of fresh semen, which is conducive for maintaining the continuity and consistency of the nerve excitement of male animals. During production, according to factors such as the age, season and body condition of the breeding stock (poultry), the semen collection operation should be strictly regulated and the semen collection frequency should be arranged reasonably.

任务实施

一、准备工作

（一）采精场地

采精场地应固定、宽敞、平坦、安静、清洁，设有假台畜或采精架（图1-6）。理想的采精场地应设有室内和室外两部分，并与精液质量检查室、输精操作室相连或距离很近。

Task Implementation

1 Preparation

1.1 Site of semen collection

It should be fixed, spacious, flat, quiet and clean, with a dummy or frame for semen collection (Figure 1-6). The ideal site of semen collection should be divided into indoor and outdoor parts, and connected or close to the semen quality inspection room and insemination room.

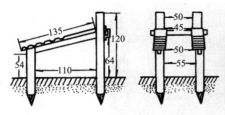

图1-6　牛的采精架（单位：cm）　　　　　采精场地

Figure 1-6　Mount stall of bull (unit: cm)　　Site of semen collection

（二）台畜

台畜有真台畜和假台畜两种。公

1.2 Dummy

There are two kinds of dummy, estrous animals

猪采精多采用假台猪（图 1-7），采精前对其彻底消毒。公牛多采用真台牛，有利于刺激公牛的性反射，也可采用假台牛（图 1-8），要求大小适宜、坚实牢固。

and fake dummy. Dummy sow is often used in semen collection(Figure 1-7). Estrous cow is usually used to stimulate the sexuality of bulls, but also can use fake dummy cow(Figure 1-8), which is required to be appropriately sized, solid and firm.

图 1-7 假台猪
Figure 1-7 Fake dummy sow

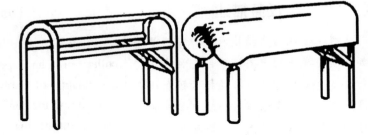

图 1-8 假台牛
Figure 1-8 Fake dummy cow

（三）器材

采精用的所有器材，均应力求清洁无菌，在使用之前要严格消毒，使用之后必须洗刷干净。

1. 假阴道（牛、羊）

（1）假阴道的结构。假阴道是模拟发情母畜阴道内环境而仿制的人工阴道，主要由外壳、内胎、集精杯、活塞等部件构成（图 1-9、图 1-10）。外壳多为硬橡胶或塑料制成，内胎为弹力强、柔软的乳胶或橡胶制成，集精杯（瓶）一般用棕色玻璃制成。

1.3 Equipment

All equipment used for sperm collection should be clean and sterile, strictly disinfected before use and cleaned after use.

1.3.1 Artificial vagina(bull, ram)

(1)Structure of artificial vagina

The artificial vagina is made by simulation the environment of estrous female's vagina. It consists of the shell, inner tube, semen collection cup, piston and so on (Figure 1-9, Figure 1-10). Its shell is mostly made of hard rubber or plastic, and the inner tube is made of elastic and soft latex or rubber. The semen collection cup is usually made of brown glass.

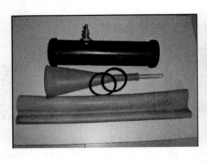

图 1-9 牛的假阴道结构
Figure 1-9 Structure of bull artificial vagina
1. 活塞 2. 外壳 3. 内胎 4. 胶圈 5. 橡胶漏斗 6. 集精杯
1. piston 2. shell 3. inner tube 4. rubber ring 5. rubber funnel 6. semen collection cup

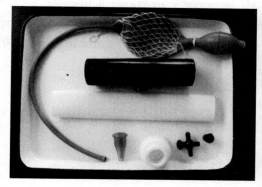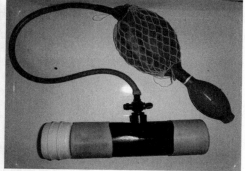

图 1-10　羊的假阴道结构
Figure 1-10　Structure of ram artificial vagina

（2）假阴道的安装。

①检查。安装前，要仔细检查外壳是否有裂口、沙眼等，内胎是否漏气、有无破损，活塞是否完好或漏气、扭动是否灵活，集精杯是否破裂等。

②安装内胎。将内胎两端翻卷于外壳上，要求松紧适度、不扭曲，内胎中轴与外壳中轴重合，即"同心圆"，再用胶圈加以固定。安装好内胎，充气调试呈 Y 形（图 1-11）。

③消毒。用长柄钳子夹取酒精棉球对集精杯消毒，同时由里向外螺旋式对内胎进行擦拭消毒。采精前，最好用生理盐水或稀释液冲洗 1~2 次，并安装集精杯。

④注水。由注水孔向外壳内注入 50~55℃ 的温水，水量为外壳与内胎容积的 1/3~1/2，主要目的是调节温度和压力。

⑤涂抹润滑剂。用消毒玻璃棒蘸取凡士林由外向内在内胎上均匀涂抹，深度为外壳长度的 1/2 左右。

⑥调压。如注入水后压力不够，可通过充气调压使假阴道入口处内胎呈 Y 形。

(2) Installation of artificial vagina

①Check. Before installation, it must be checked carefully to make sure that there are no cracks or trachoma of the shell, leaking or damage of inner tube, flaws or leaking or jam of the piston, crevices of the semen collection cup.

②Installation of inner tube. Both ends of the inner tube are rolled over the shell, making sure that the tension is moderate and there is no distortion. The inner tube axis coincides with the shell axis, and is fixed with a rubber ring. After installing the inner tube, it need to inflate to make a Y-shape(Figure 1-11).

③ Disinfection. Alcohol cotton ball is used to sterilize the inner tube and semen collection cup. Before collecting semen, it is best to rinse with saline or dilute solution 1-2 times, and then install the sterilized semen collection cup.

④ Water injection. The shell is injected with warm water at 50-55℃ by injection hole. The amount of water is about 1/3-1/2 of the volume of the shell and inner tube. The main purpose is to adjust the temperature and pressure.

⑤ Smearing lubricant. Vaseline is used to smear the inner tube from the outside to the depth with a sterile glass rod, about 1/2 of the length of the shell.

⑥Pressure adjustment. If the pressure is not high enough after water injection, it can be made into Y-shape by air at the entrance of the inner tube.

⑦ Temperature measurement. Inserting a sterilized thermometer into the cavity and reading when the temperature is constant, it is generally required to be at 38-40℃.

⑦测温。用消毒的温度计插入假阴道内腔，待温度不变时读数，一般以38～40℃为宜。

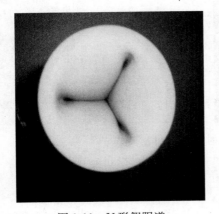

图 1-11 Y形假阴道

Figure 1-11 Y-shape artificial vagina

牛假阴道的准备

Preparation of bull artificial vagina

羊假阴道的安装

Installation of ram artificial vagina

2. 集精杯

（1）猪。将食品保鲜袋或聚乙烯袋放入采精用的保温杯中，将袋口打开，环套在保温杯口边缘，并将精液过滤纸罩在杯口上，用橡皮筋套住（图1-12），盖上盖子，放入37℃恒温箱中预热。

（2）家禽。集精杯（图1-13）必须经高压消毒后备用。集精瓶内水温应保持在30～35℃。

1.3.2 Semen collection cup

(1) Boar

Put the food preservation bag or polyethylene bag into the thermos, open the bag opening to wrap around the edge of the thermos, cover the cup rim with semen filter paper and wrap it with a rubber band (Figure 1-12), cover the lid, then preheat at a 37℃ thermostat.

(2) Poultry

The semen collection cup (Figure 1-13) must be stored after high pressure disinfection. The water temperature in the cup should be maintained at 30-35℃.

图 1-12 猪用集精杯

Figure 1-12 Semen collection cup for boar

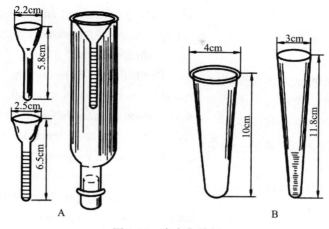

图 1-13　家禽集精杯
Figure 1-13　Semen collection cup for poultry
A. 鸡用集精杯　B. 鸭、鹅用集精杯
A. semen collection cup for cock　B. semen collection cup for drake and gander

（四）种公畜(禽)的准备

采精前，用 0.1% 高锰酸钾溶液清洗公畜包皮部并擦干。公猪要挤出包皮积尿。公鸡在采精前 3~4h 断水、断料，并剪去泄殖腔周围的羽毛。

（五）采精员的准备

采精员应技术熟练，动作敏捷，熟悉公畜的射精特点，并注意人畜安全。指甲应剪短磨光，手臂要清洗消毒等。公猪采精时，采精员要佩戴双层手套，以减少精液污染和预防人畜共患病。

二、采精操作

雄性动物的采精方法有很多，主要有假阴道法、手握法、按摩法和电刺激法等。

无论采用哪种方法采精，都要遵循以下原则：

安全：采精过程要确保操作人

1.4　Preparation of male animals

Before semen collection, the prepuce should be cleaned with 0.1% potassium permanganate solution and dried up afterwards. For boars, the accumulate urine should be squeezed out. The rooster must not be fed with any water or feed 3-4 hours before semen collection, and the feathers around the cloaca should be cut off.

1.5　Preparation of technician

The technicians should be skilled, agile, familiar with the characteristics of male ejaculation, and pay attention to the safety of human and animals. The nails should be cut short and polished, the arms should be cleaned and disinfected. In order to reduce semen contamination and prevent zoonosis, the technician must wear double-layer gloves when collecting boar semen.

2　Semen collection

There are many methods to collect sperm of male animals, such as artificial vagina method, hand holding method, massage method and electric stimulation method.

Whatever method used to collect semen, it must follow the following rules:

Security: when collecting semen, it must ensure

员、公畜等的安全，防止阴茎损伤，避免因不良刺激造成性欲下降等问题。

卫生：采精过程精液最容易受到外界因素的影响，因此，采精过程中必须小心操作，保证精液不会受到污染。

全份：必须收到全份的精液，避免精液损失。

简便：操作过程要力求简单，并且容易拆卸、清洗、消毒。

（一）公牛的采精

公牛的采精多采用假阴道法。采精时，采精员站在台牛的右侧斜后方，当公牛爬上台牛时，迅速跨前一步，左手迅速拖住包皮，右手持假阴道并调整角度使之与公牛阴茎的伸出方向呈一直线，使阴茎自然插入假阴道内（图1-14）。当公牛后肢跳起，臀部用力向前一冲，即已射精。

射精后将集精杯向下倾斜，使精液顺利流入集精杯。待阴茎自然脱离后立即竖立假阴道，打开气门，放掉空气，以充分收集滞留在假阴道内壁上的精液，然后小心取下集精杯，迅速转移至精液处理室。

the safety of operators and animals to prevent penis damage and avoid the problem of declining sexual desire due to adverse stimuli.

Health: when collecting the semen, the semen is most easily affected by external factors. Therefore, the semen collection must be handled carefully to ensure that the semen is not contaminated.

Full share: it must receive the entire amount of semen to avoid the loss of semen.

Easy: the operation process should be as simple as possible, and the equipment can be easily disassembled, cleaned and disinfected.

2.1 Bull semen collection

Artificial vagina is often used to collect bull semen. The technician stands behind the right side of the dummy cow. When the bull climbs the dummy cow, he steps forward quickly and pulls the foreskin with his left hand, holds the artificial vagina with his right hand and adjusts the angle to make it in line with the direction of bull penis. So that the penis naturally inserts into the artificial vagina(Figure 1-14). When the hind limbs of the bull jump up and the buttocks rush forward forcefully, it has ejaculated.

After ejaculation, the technician tilts the cup downward to make the semen flow into the cup smoothly. After the natural detachment of the penis, he erects the artificial vagina, opens the valve and releases the air to collect the semen that remained on the inside wall. Then the cup should be carefully removed and quickly transferred to the treatment room.

图1-14　公牛的采精

Figure 1-14　Bull semen collection

公牛采精

Bull semen collection

（二）公羊的采精

羊的采精方法与牛相似，羊从阴茎勃起到射精只有几秒，所以要求操作人员动作敏捷、准确。

采精时牵引公羊接近母羊，用发情母羊刺激公羊。采精者多站在母羊的右后方，右手持假阴道，并用食指固定集精杯，防止脱落。当公羊爬跨母羊时，迅速将假阴道口对准公羊阴茎，方向保持一致，同时用左手迅速将阴茎牵引入假阴道内，但不要触及阴茎，以免采精失败，或导致公羊恶癖。公羊有前冲动作即为射精，射精时将集精杯一端适当向下倾斜，以便精液顺利流入集精杯中。公羊射精后，待其从母羊身上退下后取出假阴道并竖立，使集精杯一端向下，放掉空气，然后取下集精杯，运往精液处理室进行精液品质检查。

公羊采精

Ram semen collection

（三）公猪的采精

目前，生产上公猪的采精方法主要有手握法和自动采精系统两种，前者具有操作简单、可选择性接取公猪精液等优点，在国内外养猪业被广泛应用；后者是一种新型的采精系统，采用仿生原理使公猪采精更接近自然状态，提高了生产效率和精液质量，适合规模化养猪场推广应用。

1. 手握法采精（图 1-15）

当公猪性欲旺盛爬跨假台猪时，采精员左手持集精杯蹲在公猪的左侧，右

2.2 Ram semen collection

The semen collection method of ram is similar to the bull. It takes only a few seconds from penile erection to ejaculation for ram, so the technician must be more agile and accurate.

When collecting semen, the technician leads the ram to the estrus ewe so as to stimulate it. The technician generally stands behind the right side of the ewe, holding the artificial vagina with the right hand, and fixing the collection cup with the index finger to prevent it from falling off. When the ram climbs across the ewe, he should quickly points the artificial vagina at the ram's penis in the same direction. At the same time, he should quickly lead the penis into the artificial vagina with the left hand, but do not touch the penis to prevent the failure of collecting semen or lead to the ram evil. The ram's forward action is ejaculation. When the ram ejaculates, the technician should tilt the collection cup to make one end down properly so that the semen will flow smoothly into the collection cup. After the ram ejaculate, they come down from the ewe. The technician should erect the artificial vagina, drop one end of the collection cup, release the air, and then take the collection cup and send it to the semen treatment room for semen quality inspection.

2.3 Boar semen collection

At present, there are two main methods for semen collection of boars: the hand holding and automatic semen collection system. The advantages of the former method are simpl and selectivity of boar semen. And it is widely used in pig husbandry worldwide. The latter is a new type of semen collection system, which uses bionic principle to make closer to the natural state. This improves the production efficiency and semen quality, and it is suitable for large-scale pig farms.

2.3.1 Hand holding method (Figure 1-15)

When the boar is sexually active and climbs across the fake dummy sow, the technician squats on

手呈锥形的空拳于公猪阴茎伸出的同时，将龟头导入空拳内，顺其向前冲力，将阴茎的S状弯曲尽可能拉直，握紧阴茎龟头防止其旋转，待充分伸展后，阴茎将停止前冲，开始射精。刚开始射出的清亮液体部分弃去不要，当射出乳白色浓精液时即可收集。公猪射精时间可持续5～10min，分2～4次射出。当公猪开始环顾四周时，说明公猪射精即将结束。采精后，小心将集精杯上的过滤砂纸及上面的胶原蛋白去掉，用盖子盖好集精杯，迅速传递到精液处理室进行检查、处理。

the left side of the boar with a cup in the left hand and a conical-shaped empty fist in the right hand. While the penis extends out, the technician leads the glans into the empty fist. Following its forward momentum, he pulls the penis as straight as possible and holds the glans tightly to prevent twisting. After fully stretching, the penis will stop rushing and begin to ejaculate. Discard the bright liquid part emitted at the beginning and collect milk-white semen. Boar ejaculation lasts for 5-10 minutes at 2-4 times. When the boars begin to look around, the ejaculation is about to end. After semen collection, remove the filter sandpaper with the collagen on it carefully and cover semen collection cup. Then transfer to the treatment room for examination and treatment quickly.

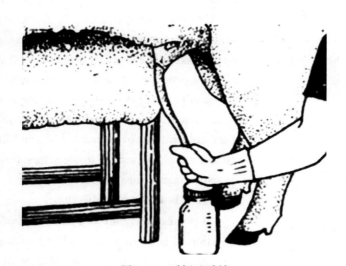

图 1-15　手握法采精　　　　　手握法采精

Figure 1-15　Hand holding method　　Hand holding method

2. 自动采精系统

自动采精系统主要由采精间、人造仿生假阴道、假台猪、主控箱组成（图1-16）。该采精系统可使公猪在射精时得到切实的生物实体感，延长种公猪的使用年限，避免精液污染，提高精液质量和生产效率。

2.3.2　Automatic semen collection system

It is mainly composed of the semen collection room, artificial bionic vagina, fake dummy sow and main control box (Figure 1-16). It can make boars get a real sense of biological entity during ejaculation. So it prolongs the service life of boars, avoids semen pollution, improves semen quality and production efficiency.

图 1-16　自动采精系统

Figure 1-16　Automatic semen collection system

自动采精系统

Automatic semen collection system

（四）公鸡的采精

公鸡的采精普遍采用背腹式按摩两人采精法（图 1-17）。采精时，一人保定公鸡，夹于腋下，双手握住两腿使其自然分叉，鸡头向后。采精员左手沿公鸡背鞍部向尾羽方向抚摸数次，以缓解公鸡惊恐并引起性兴奋，右手中指和无名指夹集精杯，杯口向外。待公鸡有性反射时，左手迅速翻转，将尾羽向背部压住，并以拇指与食指跨在泄殖腔上侧；右手拇指和食指跨在泄殖腔下侧腹部柔软部，抖动触摸数次，当泄殖腔外翻露出退化交媾器时，左手拇指与食指立刻轻轻挤压，公鸡就能排精。与此同时，迅速将集精杯口翻向泄殖腔开口处承接精液。采集的精液置于 25～30℃ 的保温瓶内以备处理。

2.4　Cock semen collection

The back-abdominal massage by two people is widely used in cock semen collection（Figure 1-17）. During semen collection, one person clamps the cock under his armpit, and grasps the legs with both hands to make it forked naturally with the cook head backward. The left hand of semen collector strokes several times along the saddle toward the tail feather to relieve its panic and cause sexual excitement. His right hand clamps the cup with the middle finger and ring finger and keep the cup rim outward. When the cock is sexually reflected, the semen collector quickly flips the left hand, presses the tail feather to the back, and cross the upper side of the cloaca with his thumb and index finger. His right thumb and index finger cross the soft part of the abdomen under the cloaca shaking and touching several times. When the cloaca turns out to reveal degraded copulatory organ, the left thumb and index finger immediately squeeze gently, and the cock can ejaculate. At the same time, quickly turn the cup to the cloaca opening to receive semen. Then put the semen in a thermos bottle at 25-30℃.

图 1-17　公鸡的采精

Figure 1-17　Cock semen collection

公鸡采精

Cock semen collection

三、采精频率

公畜（禽）的采精频率应根据其种类、个体差异、健康状况、性欲强弱、精子产生数量等确定。生产实践中，成年公牛每周 2～3 次；公羊在配种季节内可每天连续采精 2～3 次，每周 5～6 d；成年公猪隔天采精 1 次，青年公猪和老龄公猪以每周采精 2 次为宜；公鸡每周采精 4～5 次。

3 Frequency of semen collection

The semen collection frequency of male animals should be determined according to their species, individual differences, health status, sexual desire and semen production. In production practice, adult bulls can be collected semen 2-3 times a week. During mating season, the rams can be collected 2-3 times a day, 5-6 days a week. Adult boars are collected once every other day, but young and old boars are collected twice a week. For roosters, collect semen 4-5 times a week.

项目二 精液的处理
Project Ⅱ Semen Treatment

项目导学

精液的质量控制是影响人工授精效果的关键技术环节，稀释、分装、保存和运输等环节均可能会受到温度、细菌、灰尘杂质的污染而降低精液质量。精液的质量分析又易受环境、检测人员的技术熟练程度及主观判断能力等诸多因素影响而发生偏差，因此精液处理必须严格按照规范程序进行。

Project Guidance

Quality control of semen is the key links that affect the effect of artificial insemination. Dilution, packaging, preservation and transportation may be affected by temperature, bacteria and dust impurities, which may reduce the quality of semen. The quality analysis of semen is easily affected by many factors, such as environment, technical proficiency and subjective judgment of testers. Therefore, semen treatment must be carried out in strict accordance with the standard procedures.

学习目标

>>> 知识目标

- 了解精液的理化特性。
- 掌握精子的形态结构及生理特性。
- 理解外界因素对精子存活的影响。
- 理解稀释液的主要成分及作用。
- 理解精液保存的原理。

>>> 技能目标

- 能熟练评定精液的质量。
- 会确定精液的稀释倍数。
- 能熟练配制稀释液，会正确稀释精液。

Learning Objectives

>>> Knowledge Objectives

- To understand the characteristics of semen.
- To master the morphological structure and physiological characteristics of sperm.
- To understand the influence of external factors on sperm survival.
- To understand the main ingredients and functions of semen dilution.
- To understand the principles of semen preservation.

>>> Skill Objectives

- To assess semen quality.
- To calculate the multiple of semen dilution.
- To prepare the diluent and dilute semen correctly.

- 根据国家标准《牛冷冻精液》(GB/T 4143—2008)，规范制作牛的冷冻精液。
- 能正确对精液进行保存和运输。

相关知识

一、精液的组成

精液由精子和精清两部分组成。精子由睾丸产生，占的比例很小，主要化学成分为核酸、蛋白质和脂类。精清是附睾、副性腺、输精管壶腹部的分泌物，占精液的80%～90%。精液中90%～98%为水分，干物质只占2%～10%，在干物质中约有60%是蛋白质。各种动物精液量的多少，主要取决于副性腺的发达程度。牛、羊的副性腺不发达，精清分泌量少，故射精量就少，但精子密度较大；猪、马、驴的副性腺比较发达，分泌量多，其射精量也多。

不同动物精清的化学组成差异较大，即使同种动物或同一个体，因采精方法、采精时间及采精频率等不同，精清成分也有一定的变化。各种动物的精清成分见表2-1。

- Produce frozen bovine semen according to national standard "Frozen bovine semen"(GB/T 4143—2008).
- To preserve and transport the semen correctly.

Relevant Knowledge

1 Composition of semen

Semen consists of sperm and seminal plasma. Sperm is produced by testis, which accounts for a small proportion of semen. The main chemical constituents of sperm are nucleic acids, proteins and lipids. Seminal plasma is the secretion of epididymis, accessory gonads and vas deferens ampulla, accounting for 80%-90% of semen. 90%-98% of semen is water, and only 2%-10% of semen is dry matter, 60% of dry matter is protein. The amount of semen in animals depends mainly on the development of accessory gonads. The accessory gonads of cattle and sheep are less developed, so the secretion of seminal plasma is less than that of pig and horse. Therefore, the volume of sperm in cattle and sheep is small, but the density of sperm is large. The accessory gonads of pig, horse and donkey are highly developed, thus having more seminal plasma and more semen.

The chemical composition of seminal plasma in different animals is quite different. Even for the same species or the same individual, due to different methods, time and frequency of semen collection, the seminal plasma ingredients also undergo certain changes. The ingredients of seminal plasma in various animals are shown in Table 2-1.

表2-1 各种动物精清的化学组成
Table 2-1 The components of seminal plasma

成分 Components	动物种类 Animal species			
	牛 Cattle	羊 Sheep/Goat	猪 Pig	鸡 Chicken
蛋白质 Protein/(g/dL)	6.8	5.0	3.7	1.2～2.8

（续）

成分 Components	动物种类 Animal species			
	牛 Cattle	羊 Sheep/Goat	猪 Pig	鸡 Chicken
果糖 Fructose/（mg/dL）	460～600	250	9	4
山梨醇 Sorbitol/（mg/dL）	10～140	26～170	6～8	0～10
肌醇 Inositol/（mg/dL）	25～46	7～14	380～630	16～20
柠檬酸 Citric acid/（mg/dL）	620～806	110～260	173	0
甘油磷脂酰胆碱 Glycerol phosphatidylcholine/（mg/dL）	100～500	1 100～2 100	110～240	0～40
钠 Sodium/（mg/dL）	225±13	178±11	587	352
钾 Potassium/（mg/dL）	155±6	89±4	197	61
钙 Calcium/（mg/dL）	40±2	6±2	6	10
镁 Magnesium/（mg/dL）	8±0.3	6±0.8	5～14	14
氯化物 Chloride/（mg/dL）	174～320	86	260～430	147

二、精子的形态结构

各种动物精子的形状、大小及内部结构有所不同，但大体上是相似的（图 2-1）。哺乳动物的精子整体形状呈蝌蚪状，主要由头部、颈部和尾部组成（图 2-2）。精子长度因动物种类不同而有差异，家畜精子的长度为 50～90 μm。精子的长度和体积与动物自身大小无关，如大鼠精子长约 190 μm，而大象的精子只有 50 μm 长。

2 Morphological structure of sperm

The shape, size and internal structure of sperm of various animals are different, but they are generally similar(Figure 2-1).The sperm of mammals are tadpole-shaped, consisting of head, neck and tail(Figure 2-2). Sperm length varies according to animal species. Sperm length of livestock is about 50-90 μm. The length and volume of sperm have no relevance to the size of the animal. For example, the sperm of the mouse is about 190 μm long, while the sperm of the elephant is only 50 μm long.

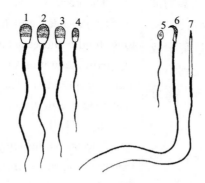

图 2-1 各种动物精子的形态
Figure 2-1 Morphology of sperms in various animals
1. 牛 2. 猪 3. 羊 4. 马 5. 人 6. 小鼠 7. 鸡
1. bull 2. boar 3. ram 4. horse 5. human 6. mouse 7. rooster

图 2-2 精子的形态结构
Figure 2-2 Morphological structure of sperm
1. 头 2. 颈 3. 顶体 4. 中段 5. 主段 6. 末段
1. head 2. neck 3. acrosome 4. middle piece 5. principal piece 6. end piece

1. 头部

精子的头部主要由细胞核和顶体构成，核内含有遗传物质 DNA。核的前端为顶体，是一个由双层膜组成的帽状结构，覆盖在核的前 2/3 部分，靠近质膜的一层称为顶体外膜，靠近核的一层称为顶体内膜。顶体内含有多种与受精有关的酶，是一个不稳定的特殊结构，其畸形、缺损或脱落会使精子的受精能力降低或完全丧失。精子顶体异常率高低是评定精液品质好坏的重要指标之一。

2. 颈部

精子的颈部是连接头部和尾部的部分，呈短圆柱状。精子颈部长约 0.5 μm，脆弱易断，特别是在精子成熟过程中或在精液稀释、保存、运输时，受到不良影响，极易从颈部断开，成为头、尾分离的畸形精子。

3. 尾部

尾部为精子最长的部分，是精子的代谢器官和运动器官。尾部长 40～50 μm，分为中段、主段及末段三部分。精子主要靠尾部鞭索状波动推动精子向前运动。

2.1 Head

The head of sperm is mainly composed of nucleus and acrosome, and the nucleus contains DNA. The front end of the nucleus is the acrosome, which is a cap-like structure consisting of two layers of membranes covering the first two-thirds of the nucleus. The layer close to the plasma membrane is called the outer acrosome membrane, and the layer close to the nucleus is called the inner acrosome membrane. The acrosome contains a variety of enzymes related to fertilization, which is an unstable structure. The abnormality and defects of acrosome can reduce or completely lose the fertilization ability of sperm. Sperm acrosome abnormality rate is one of the important indicators to evaluate the quality of semen.

2.2 Neck

The neck of sperm is a short cylindrical part connecting the head and tail. The neck is about 0.5 μm long, fragile and easy to break, especially in the process of sperm maturation or in semen dilution, preservation and transportation. The fracture usually leads to a deformed sperm disconnecting from the neck.

2.3 Tail

The tail is the longest part of sperm, which is the metabolic organ and motor organ. The tail length ranges from 40 μm to 50 μm, which is divided into three parts: middle piece, principal piece and end piece. Sperm is mainly driven forward by whiplash-like fluctuations of the tail.

三、 精子的生理特性

(一) 精子的代谢

新陈代谢是精子维持其生命和运动能力的基础。精子代谢主要是通过糖酵解和呼吸作用两种方式进行，这是在不同条件下既有联系又有区别的代谢过程。

1. 糖酵解

糖类是精子代谢的主要基质。无论在有氧还是无氧条件下，精子都可以把精清中的果糖或稀释液中的葡萄糖、乳糖、蔗糖等物质分解成乳酸而释放出能量，此过程称为糖酵解。

2. 呼吸作用

在有氧条件下，精子可将糖酵解过程中产生的乳酸、丙酮酸等有机酸，通过三羧酸循环彻底分解为 CO_2 和水，产生更多的能量。呼吸旺盛，会使氧和代谢基质消耗过快，造成精子早衰，所以在保存精液时应采取隔绝空气或充入 CO_2、降低温度及 pH 等办法，尽量减少能量消耗，延长其体外存活时间。

(二) 精子的运动

运动能力是精子有生命力的重要特征之一。精子的运动依赖于尾部的摆动。显微镜下观察到的精子运动类型有三种（图2-3）：直线前进运动、转圈运动和原地摆动。其中，直线前进运动是精子正常的运动形式，这样的精子能运行到受精部位参与受精作用，称为有效精子。而转圈运动和原地摆动的精子不具备受精能力。

3 Physiological characteristics of sperm

3.1 Sperm metabolism

Metabolism is the basis for sperm to maintain its life and motor ability. Sperm metabolism is mainly carried out by glycolysis and respiration, which are both related and different metabolic processes under different conditions.

3.1.1 Glycolysis

Carbohydrates are the main substrates of sperm metabolism. Under both aerobic and anaerobic conditions, sperm can decompose fructose in the semen or glucose, lactose and sucrose in dilution into lactic acid to release energy. This process is called glycolysis.

3.1.2 Respiration

Under aerobic conditions, sperm can completely decompose lactic acid, pyruvate and other organic acids produced during glycolysis into carbon dioxide and water through tricarboxylic acid cycle, generating more energy. Vigorous breathing will cause excessive consumption of oxygen and metabolic matrix, resulting in premature sperm death. Therefore, when storing semen, it should be isolated from air or filled with carbon dioxide, with lower temperature and pH, to minimize energy consumption and prolong its survival time in vitro.

3.2 Sperm motility

Motor ability is one of the important indicators of sperm vitality. Sperm movement depends on the tail swing. There are three types of sperm motions observed under a microscope, namely straight forward motion, circle motion, and swing in situ (Figure 2-3). Among them, straight forward motion is the normal form of sperm movement, such sperm can run to the fertilization site to participate in fertilization, called effective sperm. The sperm that moves in circles and swings in situ do not have the ability to fertilize.

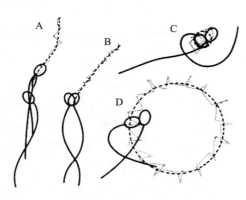

图 2-3 精子运动轨迹
Figure 2-3 Motion trajectory of sperm
A、B. 直线前进运动　C. 原地摆动　D. 转圈运动
A，B. straight forward motion　C. swing in situ　D. circle motion

精子的运动方式
Form of sperm movement

四、外界因素对精子的影响

影响精子生存的外界因素很多，如温度、光照和辐射、渗透压、pH、电解质浓度、精液的稀释及药品等。在生产中，工作人员要充分了解这些因素的影响，才能在精液的稀释和保存中，控制适宜的条件，延长精子存活和保持受精能力的时间，不断提高人工授精技术的应用效果。

1. 温度

精子对温度十分敏感，体温是保持精子正常代谢和运动的最适温度，哺乳动物的体温为 37~38℃，禽类为 40℃。但在这种温度下，不利于精子的长时间保存。精子对高温的耐受性差，一般不超过 45℃。当温度超过这一限度时，精子经过短促的热僵直后迅速死亡。在低温环境下，精子的代谢和运动受到抑制，能量消耗减少。若精液由体温急剧降至 10℃ 以下时，精子受到冷打击，不可逆地失去活力

4 Influence of external factors on sperm

Many external factors affect sperm's survival, such as temperature, light and radiation, osmotic pressure, pH, electrolyte concentration, semen dilution and medicine etc. In production process, the staff should fully understand the impact of these factors in order to control the appropriate conditions in the dilution and preservation of semen, prolong the time of sperm survival and fertilization ability, and continuously improve the efficiency of artificial insemination.

4.1 Temperature

Sperm is very sensitive to temperature. Body temperature is the optimum temperature to maintain normal sperm metabolism and movement. The body temperature of mammals is between 37-38℃, and that of poultry is 40℃. However, at this temperature, it is not conducive to the long-term preservation of sperm. The tolerance of sperm to high temperature is poor, usually no higher than 45℃. When the temperature exceeds this limit, the sperm dies quickly after a short thermal stiffness. At low temperatures, sperm metabolism and movement are inhibited, its energy consumption is reduced. If the temperature drops sharply from body

而不能复苏,这种现象称为冷休克。在超低温(-196℃)环境中,精子的代谢和运动基本停止,生命处于静止状态,可进行长期保存。解冻后,精子活力复苏且不丧失受精能力。

2. 光照和辐射

可见光、紫外光及各种放射性射线均会对精子的活力产生不利影响。直射的日光可提升精子的摄取氧能力,会加速精子的呼吸和运动,从而缩短精子寿命。紫外线对精子的影响取决于它的强度,其中波长 240 nm 的紫外光对精子的危害最大。荧光灯发射的紫外线,对精子也有不利影响。因此,在采精和精液处理时,应尽量减少光的照射,装精液的容器最好采用棕色瓶。

3. 渗透压

精子只有与其周围的液体(如稀释液)保持等渗,才能维持细胞的完整性和正常的生理活动。周围液体的渗透压过高时,精子内的水分向外渗出,精子会失水皱缩,严重时会死亡;而在低渗环境中,水分向精子内渗透,使精子膨胀、细胞膜破裂而导致精子死亡。

4. pH

精液 pH 的变化可明显地影响精子的代谢和活动力,pH 可用 pH 试纸测定。在弱酸性环境中,精子的代谢和活动受到抑制;反之,当 pH 升高时,精子代谢和呼吸增强,运动和能量消耗加剧,精子寿命相对缩短。因此,弱酸性环境更有利于精液保存,可通过向精

temperature to 10℃, the sperm will be damaged by the cold environment, irreversibly lose vitality and can not recover. This phenomenon is called "cold shock". In ultralow temperature (-196℃) environment, sperm metabolism and movement are almost stopped, its life is in a static state, and it can be preserved for a long time. After thawing, sperm motility is restored without loss of fertilization capacity.

4.2　Light and radiation

Visible light, ultraviolet light and various radioactive rays can adversely affect the vitality of sperm. Sunlight can improve the oxygen uptake of sperm, accelerate the breathing and movement, and shorten its life. The effect of ultraviolet light on sperm depends on its intensity, and ultraviolet light with a wavelength of 240 nm is the most harmful. Ultraviolet light emitted by fluorescent lamps also has an adverse effect on sperm. Therefore, in the process of semen collection and semen treatment, the irradiation of light should be minimized, and the container for the semen is best to use brown bottles.

4.3　Osmotic pressure

Sperm can only maintain its cell integrity and normal physiological activities when it remains isotonic with its surrounding fluid (such as diluent). When the osmotic pressure of the surrounding liquid is too high, the sperm will lose water and shrink, and will die in severe cases. In the hypotonic environment, the water will penetrate into the sperm, causing the swell and cell membrane rupture, leading sperm to death.

4.4　pH

Changes in semen pH can significantly affect sperm metabolism and motility, and pH can be measured using pH paper. In a weakly acidic environment, sperm metabolism and activity is inhibited. Conversely, when pH rises, sperm metabolism and respiration is enhanced, movement and energy consumption increased, so the sperm life is relatively shortened. Therefore, a weakly acidic environment is more conducive to semen preservation, and the pH can be

中充入饱和 CO_2 气体或使用碳酸盐等方法降低 pH。

精子适宜的 pH 范围因动物种类不同而有差异，一般为：牛 6.9~7.0，羊 7.0~7.2，猪 7.2~7.5，家兔 6.8，鸡 7.3。

5. 电解质浓度

精子的代谢和活力受环境离子类型和浓度的影响。一定量的电解质对精子的正常刺激和代谢是必要的，尤其是一些弱碱性盐类，如柠檬酸盐、磷酸盐等溶液，对维持精液 pH 的相对稳定具有重要作用。但高浓度的电解质易破坏精子与精清的等渗性，易对精子造成损害。

6. 精液的稀释

精液经过稀释处理后，精子的代谢和运动加强，受精能力会增强。在进行高倍稀释时，会使精子膜通透性增大，影响精子的代谢和生存。在稀释液中加入卵黄并且采用分步稀释的方式，可减少高倍稀释对精子的有害影响。稀释精液时，应该将稀释液沿着容器壁缓缓加入精液中，边加入边摇晃，以防稀释打击。

7. 药品

一些药品对精子具有保护作用。例如，向精液或稀释液中加入适量抗生素、磺胺类药物，能抑制病原微生物的繁殖，有利于精子的保存；在稀释液中加入适量甘油、二甲基亚砜，可缓解冷冻过程对精子的伤害。但是，某些防腐消毒药品，如酒精、煤皂酚等对精子有害。所以，在人工授精操作中，工作人

reduced by filling semen with saturated carbon dioxide or using carbonate method etc.

The appropriate pH range of sperm varies depending on the animal species, generally, 6.9-7.0 for bull, 7.0-7.2 for ram, 7.2-7.5 for boar, 6.8 for male rabbit, 7.3 for cock.

4.5 Electrolyte concentration

The metabolism and motility of sperm are affected by the type and concentration of environmental ions. A certain amount of electrolytes is necessary for the normal stimulation and metabolism of sperm, especially some weak alkaline salts, such as citrate, phosphate and other solutions, which play an important role in maintaining the stability of semen pH. However, the high concentration of electrolyte will easily destroy the isotonicity of sperm and seminal plasma, and it is easy to cause damage to sperm.

4.6 Dilution of semen

After dilution, the metabolism and movement of the sperm are strengthened, and its fertilization ability is enhanced. When high-dilution is performed, the permeability of the sperm membrane is increased, which affects sperm metabolism and survival. The harmful effects of high-dilution on sperm can be reduced by adding egg yolk into the dilution solution and adopting a step-by-step dilution. When the semen is diluted, the dilution should be slowly added to the semen along the wall of the container and shaken while adding to reduce the direct impact of dilution.

4.7 Medicine

Some medicines have protective effects on sperm. For example, adding an appropriate amount of antibiotics or sulfonamides to semen or diluent can inhibit the reproduction of pathogenic microorganisms and facilitate sperm preservation, adding appropriate amount of glycerin and dimethyl sulfoxide to diluent can alleviate the damage to sperm in freezing process. However, certain antiseptic and disinfecting drugs, such as alcohol and lysol, are harmful to sperm. Therefore, in

员既要保持所用器械清洁无菌,又要避免将消毒药液混入精液中。

五、精液的稀释

在精液中添加一定数量的、适宜于精子存活并保持其受精能力的溶液,称为精液的稀释。精液稀释的目的在于扩大精液容量、提高受配母畜头数、延长精子的存活时间和受精能力、利于精液的保存与运输。

(一)稀释液的成分及作用

1. 稀释剂

稀释剂主要用于扩大精液的容量,要求稀释剂与精液具有相同的渗透压。多采用等渗生理盐水、5%葡萄糖等。

2. 营养剂

营养剂主要是为精子提供营养,补充精子生存和运动所消耗的能量。常用的营养剂有糖类、卵黄、乳类等。

3. 保护剂

(1)缓冲物质。用以保持精液适当的pH。精液在保存过程中,随着代谢产物(乳酸和CO_2等)的累积,精液的pH会逐渐下降,下降到一定程度时,精子甚至会发生酸中毒,不可逆地失去活力。因此,向精液中添加一定量的缓冲物质,可使pH稳定在一定范围之内。常用的无机缓冲剂主要有柠檬酸钠、磷酸二氢钾、酒石酸钾钠等;有机缓冲剂有三羟甲基氨基甲烷(Tris)和乙二胺四乙酸二钠(EDTA)。

(2)防冷物质。具有防止精子冷休克的作用。在精液保存过程中常需要降

the artificial insemination operation, the operator must keep the instruments used clean and sterile, and avoid the mixing-up of disinfectant liquid and semen.

5 Semen dilution

Adding a certain amount of solution in the semen, which is suitable for sperm survival and maintaining fertilization ability is called semen dilution. The purpose of semen dilution is to expand the semen volume, increase the number of mated females, prolong the survival time and fertilization ability of sperm, and facilitate the preservation and transportation of semen.

5.1 Composition and functions of diluent

5.1.1 Dilutant

It is mainly used for enlarging the volume of semen, requiring the same osmotic pressure between dilutant and semen. Isotopic physiological saline and 5% glucose are often used.

5.1.2 Nutrients

It mainly provides nutrition for sperm and supplements the energy consumed by its survival and movement. Sugar, egg yolk and milk are usually used as nutrients.

5.1.3 Protective agents

5.1.3.1 Buffer substances

Buffer substances are used to maintain the proper pH of semen. In the process of semen preservation, with the accumulation of metabolites(lactic acid and carbon dioxide, etc.), the pH of semen will gradually decrease, and when it drops to a certain extent, the sperm will even be acidosis, which makes the sperm irreversibly lose its vitality. Therefore, adding a certain amount of buffer substances to semen can stabilize the pH within a certain range. The common inorganic buffers are sodium citrate, potassium dihydrogen phosphate, potassium sodium tartrate, etc. The organic buffers are Tris and EDTA.

5.1.3.2 Anti-cold substances

It has the effect of preventing sperm from cold

温处理,尤其是从 30℃ 急剧下降到 10℃ 时,因精子内含有的缩醛磷脂在低温下容易凝固,影响精子的正常代谢,造成精子不可逆的冷休克而丧失活力。卵黄和乳类含有卵磷脂,其熔点低,在低温下不易被冻结,可防止精子冷休克而起到保护作用。

(3) 抗冻物质。可以防止冷冻过程中"冰晶化"。精液在冷冻和解冻过程中,精液所经历的固、液态之间的转化对精子的危害很大。甘油、二甲基亚砜可有效缓解这种危害,是生产中常用的抗冻保护剂。

(4) 抗菌物质。具有抗菌作用。在采精和精液处理过程中,向稀释液中加入一定量的抗生素可以防止精液遭受微生物的污染。常用的抗生素有青霉素、链霉素、林可霉素、卡那霉素、恩诺沙星。

(5) 非电解质和弱电解质。向精液中添加适量的非电解质或弱电解质物质,具有降低精清中电解质浓度的作用。常用的非电解质和弱电解质有各种糖类、氨基乙酸等。

4. 其他添加剂

这类添加剂主要用于改善精子外在环境的理化特性,以及母畜生殖道的生理机能,从而提高受胎率。常用的有酶类、激素类和维生素类物质。如 β-淀粉酶、催产素、前列腺素、维生素 B_1、维生素 B_2、维生素 B_{12}、维生素 C 等。

shock. The cooling process is usually required during semen preservation. Once the temperature drops from 30℃ to 10℃ sharply, the plasmalogen contained in sperm easily condenses at this process, affecting the normal metabolism of sperm, causing the sperm to be irreversibly cold shocked and lose its vitality. Egg yolk and milk contain lecithin, which has a low melting point and is not easily frozen at low temperatures can play a protective role in protecting sperm from cold shock.

5.1.3.3　Antifreeze substances

It is possible to prevent the formation of ice crystallization during the freezing process. During the freezing and thawing process of semen, the conversion between the solid and liquid phases is very harmful to the sperm. Glycerin and dimethyl sulfoxide can effectively alleviate this hazard and are commonly used as antifreeze protectants in production.

5.1.3.4　Antibacterial substances

It has antimicrobial effect. In the process of semen collection and semen treatment, adding a certain amount of antibiotics to the diluent can prevent semen from being contaminated by microorganisms. The commonly used antibiotics are penicillin, streptomycin, lincomycin, kanamycin and enrofloxacin.

5.1.3.5　Non-electrolytes and weak electrolytes

Adding appropriate amount of non-electrolytes or weak electrolyte substances to semen can reduce the concentration of electrolytes in semen. The commonly used non-electrolytes and weak electrolytes are various sugars, aminoacetic acids, etc.

5.1.4　Other additives

These additives are mainly used to improve the physical and chemical properties of the external environment of sperm, as well as the physiological functions of the genital tract of female animal, so as to improve the conception rate. The commonly used additives are enzymes, hormones and vitamins. For example, β-amylase, oxytocin, prostaglandin, vitamin B_1, vitamin B_2, vitamin B_{12}, vitamin C, etc.

（二）稀释液的种类

1. 现用稀释液

现用稀释液适用于精液稀释后立即输精用，不进行保存。以单纯扩大精液容量、增加输精母畜头数为目的。此类稀释液常以简单的等渗糖类和乳类溶液为主，也可选用生理盐水。

2. 常温保存稀释液

常温保存稀释液适用于精液的常温短期保存。以糖类和弱酸盐为主体，一般 pH 控制在 6.35 左右。

3. 低温保存稀释液

低温保存稀释液适用于精液的低温保存。以卵黄和乳类为主体，具有抗冷休克的作用。

4. 冷冻保存稀释液

冷冻保存稀释液适用于精液冷冻保存。其稀释液成分较为复杂，除糖类、卵黄外，还应添加甘油或二甲基亚砜抗冻剂。

生产中选用稀释液时，应根据用途、动物种类及精液的保存时间、保存方法等进行综合考虑，选择来源广、成本低、效果好、易配制的稀释液配方。

（三）稀释倍数

精液进行适当的稀释可以提高精子的存活率，但如果稀释倍数超过一定限度，其存活率就会受到影响。精液稀释倍数应根据精液的质量来确定，尤其是精子的活力和密度，还有每次输精所需要的有效精子数、稀释液的种类和保存方法等。

（1）牛、羊精子密度大，稀释倍数可大一点，而猪、马精子密度小，不宜

5.2 Types of diluents

5.2.1 Simple diluent

It is suitable for insemination immediately after semen dilution and is not used for preservation. Using this type of diluent is to expand the semen volume and increase the number of inseminated female animals. Simple isotonic sugars and dairy solutions are often used in the preparation of such diluents. Normal saline can also be used.

5.2.2 Preservation diluent at room temperature

It is suitable for short-term storage of semen at room temperature. The main components are sugars and weak acid salts, and the pH is generally controlled at about 6.35.

5.2.3 Preservation diluent at low temperature

It is suitable for cryopreservation of semen. Egg yolk and milk are the main components, which have the function of anti-cold shock.

5.2.4 Cryopreservation dilution

It is suitable for cryopreservation of semen. The composition of the diluent is complicated, and in addition to sugars and egg yolk, glycerin or dimethyl sulfoxide antifreeze should be added.

When choosing a diluent in production, comprehensive consideration should be made according to the purpose, animal species, storage time and storage method of semen, etc., and a diluent formula with a wide range of sources, low cost, good effect, and easy preparation should be selected.

5.3 Dilution ratio

Proper dilution of semen can improve the survival rate of sperm, but if the dilution ratio exceeds a certain limit, the survival rate will be affected. The dilution ratio should be determined according to the quality of semen, especially the motility and density of sperm, the number of effective sperm needed for each insemination, as well as the type of diluent and the method of preservation.

(1) Sperm density of bull and ram is large, and it can be highly diluted, while boar and male horse is

作高倍稀释。

（2）乳类稀释液可作高倍稀释，糖类稀释液不宜作高倍稀释。

（3）冷冻保存稀释倍数应低一些，液态保存稀释倍数可高些。

牛精液耐稀释的潜力很大，但生产上仅稀释 10～40 倍。山羊与牛相似。绵羊精液稀释后 1h，受精率就会有所下降，因此常不作稀释即用于输精。公猪精液一般稀释 2～4 倍。

六、精液保存的原理

精液保存的目的是延长精子在体外存活的时间，便于长途运输，从而扩大精液的使用范围。精液保存的方法，按保存温度不同可分为常温（15～25℃）保存、低温（0～5℃）保存和冷冻（－196℃）保存三种。

1. 常温保存的原理

常温保存又称为室温保存或变温保存。常温保存不需要特殊设备，简单易行，便于普及和推广，适用于各种动物精液的短期保存，尤其适合公猪精液的保存。其保存原理是：精子在弱酸性环境中，其活动受到抑制，能量消耗减少，而当 pH 恢复到中性，精子活力即可复苏。因此，可在稀释液中加入弱酸性物质（如己酸），把 pH 调整到 6.35 左右，从而抑制精子的活动。亦可向稀释液内充入一定量的 CO_2 气体，溶入水形成碳酸，变成弱酸性环境，达到短期保存精液的目的。不同酸类物质对精子产生的抑制区域和保护效果不同，一般认为有机酸好于无机酸。

small in density and not suitable for high dilution.

(2) Milk diluents are suitable for high dilution, while sugar diluents are not suitable for high dilution.

(3) The dilution ratio of cryopreservation should be lower and that of liquid preservation could be higher.

Bull semen has a great potential for dilution resistance, but it is only 10-40 times diluted in production. Goats are the same as cattle in that respect. The fertilization rate of sheep decreases in one hour after diluting, so it is not suitable for dilution. Boar semen is generally diluted 2-4 times.

6 Principle of semen preservation

The purpose of semen preservation is to prolong the survival time of sperm in vitro, facilitate long-distance transportation, and expand the range of application. According to the storage temperature, it can be divided into normal temperature preservation (15-25℃), low-temperature preservation (0-5℃) and cryopreservation (－196℃).

6.1 Principle of normal temperature preservation

The normal temperature preservation is also called room temperature or variable temperature preservation. It is simple and easy to be popularized. It is suitable for short-term preservation of animal semen, especially boar semen. The principle of semen preservation is that sperm activity is inhibited and energy consumption is reduced in weak acid environment, and sperm motility can be recovered once the pH is returned to neutral. Therefore, weak acid substances (such as hexanoic acid) can be added into the diluent to adjust the pH to about 6.35, so as to inhibit sperm activity. Moderate carbon dioxide is filled into the diluent to dissolve into water to form carbonic acid, which will create a weak acid environment, so as to achieve the purpose of short-term preservation of semen. Different acids have different inhibitory regions and protective effects on sperm.

2. 低温保存的原理

低温保存是将稀释后的精液置于 0~5℃的环境中保存，保存时间通常比常温保存时间长，但公猪的精液不如常温保存效果好。低温保存原理为：当温度缓慢降至 0~5℃时，精子呈现"休眠"状态，精子代谢机能和活动力减弱，当温度回升后，精子又逐渐恢复正常的代谢机能而不丧失其受精能力。

3. 冷冻保存的原理

精液冷冻保存是用液氮（-196℃）或干冰（-79℃）作冷源，达到长期保存的目的。该方法保存时间长，精液的使用不受时间和地域的限制，是一种比较理想的保存方法，对人工授精技术的推广及现代畜牧业的发展都具有十分重要的意义。目前，牛的冷冻精液在生产上的普及率已达到100%。其保存原理为：在超低温下，精子运动和代谢完全停止，生命以"静止"状态保存下来，当温度回升后，又能复苏且不丧失受精能力。但精子的复苏率只有 50%~70%，部分精子在冷冻过程中死亡。

七、液氮的特性及液氮罐的结构

（一）液氮的特性

液氮是空气中的氮气经分离、压缩形成的一种无色、无味、无毒的透明液体，沸点为-195.8℃，每升液氮为0.8kg左右。液氮具有超低温性，可抑制精子代谢和细菌繁殖，能长期保存精液；液氮具有很强的膨胀性，当温度达到15℃时，在标准大气压下，1L液氮可气化为

6.2 Principle of low-temperature preservation

Low-temperature preservation is to store the diluted semen in the environment of 0-5℃. The preservation time is usually longer than that at normal temperature, but effect of boar semen is not as good as that of normal temperature. The principle of low-temperature preservation is as follows: when the temperature drops slowly to 0-5℃, the sperm presents a "dormant" state, and its metabolic function and activity are weakened. When the temperature rises, the sperm gradually returns to normal metabolic function without losing fertilization ability.

6.3 Principle of cryopreservation

Semen cryopreservation is to use liquid nitrogen (-196℃) or dry ice (-79℃) as cold source to achieve the purpose of long-term preservation. It is an ideal method to preserve semen for a long time. It is of great significance to the popularization of artificial insemination technology and the development of modern animal husbandry. At present, the popularization rate of frozen semen in cattle has reached 100%. The principle of cryopreservation is as follows: under ultra-low temperature, sperm movement and metabolism are completely stopped, life is preserved in a "static" state, and when the temperature rises, it can recover without losing fertilization ability. However, the recovery rate of sperm was only 50%-70%, and some sperm died during freezing.

7 Characteristics of liquid nitrogen and structure of liquid nitrogen tank

7.1 Characteristics of liquid nitrogen

Liquid nitrogen is a colorless, odorless and nontoxic transparent liquid formed by the separation and compression of nitrogen in the air. The boiling point is -195.8℃, and the weight of liquid nitrogen is about 0.8 kg per liter.

Liquid nitrogen has ultra-low temperature, which can inhibit sperm metabolism and bacteria reproduction, and can preserve semen for a long time. Liquid

680L氮气，膨胀率为680倍；液氮具有挥发性，当液氮用量大时，要注意通风，以防窒息。

nitrogen has strong expansibility, under standard atmospheric pressure, when the temperature reaches 15℃, one liter liquid nitrogen can be gasified into 680 liters nitrogen. Liquid nitrogen is volatile, when the amount of liquid nitrogen is large, pay attention to ventilation to prevent suffocation.

（二）液氮罐的结构

液氮罐可分为贮存罐和运输罐两种。贮存罐主要用于室内液氮的静置贮存，容量大小不等，大的可达数百升，小的不到1L，一般的人工授精站适宜用10～30 L的中型罐。为了满足运输条件，液氮运输罐增加了专门的防震设计，但也应避免剧烈的碰撞和震动。

液氮罐由外壳、内槽、夹层、颈管、盖塞、提筒及外套构成（图2-4）。

7.2 Structure of liquid nitrogen tank

There are two types of nitrogen tanks, storage tank and transport tank. The storage tank is mainly used for the static storage of liquid nitrogen indoors. The capacity varies from less than 1 liter to several hundred liters. It is suitable for general artificial insemination station to use 10-30 liters medium-sized tank. In order to meet the transportation condition, the special shock proof design is added to the transport tank, but violent collision and vibration should be avoided.

The liquid nitrogen tank is composed of shell, inner groove, interlayer, neck tube, cap plug, lifting cylinder and coat(Figure 2-4).

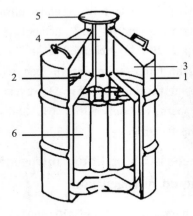

图 2-4 液氮罐
Figure 2-4 The liquid nitrogen tank
1. 外壳 2. 内槽 3. 夹层 4. 颈管 5. 盖塞 6. 提筒
1. shell 2. inner groove 3. interlayer 4. neck tube 5. cap plug 6. lifting cylinder

液氮罐结构
Structure of liquid nitrogen tank

1. 外壳

液氮罐的罐壁由内、外两层构成，外层称为外壳，内层称为内胆，一般由

7.2.1 Shell

The wall of liquid nitrogen tank is composed of inner and outer layers. The outer layer is called the

坚硬的合金制成。

2. 内槽

液氮罐内层中的空间称为内槽。内槽的底部有底座,供固定提筒用,可将液氮及冷冻精液储存于内槽中。

3. 夹层

内外两层间的空隙为夹层。为增加罐的保温性,夹层被抽成真空,在夹层中装有绝热材料和吸附剂(如活性炭),以吸收漏入夹层的空气,从而增加了罐的绝热性能。

4. 颈管

颈管以绝热黏合剂将罐的内外两层连接,并保持有一定的长度。顶部有罐口,其结构既要有孔隙能排出液氮蒸发出来的氮气以保证安全,又要有绝热性能以尽量减少液氮的气化量。

5. 盖塞

盖塞由绝热性能良好的塑料制成,以阻止液氮的蒸发,又可固定贮精提筒的手柄。

6. 提筒

提筒是存放冻精的装置。提筒的手柄由绝热性能良好的塑料制成,既能防止热量向液氮传导,又能避免取冻精时冻伤。提筒的底部有多个小孔,以便液氮渗入其中。

7. 外套

中、小型液氮罐为了携带方便,有

shell, and the inner layer is called the liner. It is generally made of hard alloy.

7.2.2　Inner groove

The space in the inner layer of the liquid nitrogen tank is called the inner groove. There is a base at the bottom of the inner tank for fixing the lifting cylinder, which can store liquid nitrogen and frozen semen in the inner tank.

7.2.3　Interlayer

The gap between the inner and outer layers is the interlayer. In order to improve the thermal insulation of the tank, the interlayer is pumped into a vacuum, the insulation material and adsorbent (such as activated carbon) are installed in the interlayer to absorb the air leaking into the interlayer, thus increasing the thermal-insulation performance of the tank.

7.2.4　Neck tube

The neck tube connects the inner and outer layers of the tank with insulating adhesive and keeps a certain length. There is an opening on the top of the tank. The structure should not only have holes to discharge the nitrogen to ensure safety, but also have insulation performance to reduce the gasification amount of liquid nitrogen as much as possible.

7.2.5　Cap plug

The cap plug is made of plastic with good insulation performance to prevent the evaporation of liquid nitrogen, and can fix the handle of lifting cylinder.

7.2.6　Lifting cylinder

The lifting cylinder is a device for storing frozen semen. The handle of lifting cylinder is made of plastic with good insulation performance, which can not only prevent the heat from transmitting to liquid nitrogen, but also avoid frostbite when taking frozen semen. There are many small holes at the bottom of the lifting cylinder so that liquid nitrogen can penetrate into it.

7.2.7　Coat

For the convenience of carrying, the small and

一外套并附有挎背用的皮带。

液氮罐都应该具有绝热性能好、坚固耐用、使用方便等特点，这样才能更好地满足生产的需要。

（三）液氮罐的使用

1. 使用前要认真检查

新购或长期未用的液氮罐，必须外部无破损、无异常，内部干燥无异物，颈管和盖塞完全，贮精提筒完好，盛装液氮经过 1d 预冷并观察其损耗率，各项指标合格后方可使用。

2. 填充液氮时要小心谨慎

对于新罐或处于干燥状态的罐一定要缓慢填充并进行预冷，以防降温太快损坏内胆，减少使用年限。或者将液氮运输罐的液氮经漏斗注入贮存罐内，为了防止液氮飞溅，可在漏斗内衬一块纱布。

3. 及时补充液氮

当液氮消耗掉 1/2 时，应立即补充液氮。罐内液氮的剩余量可用称量法来估算，也可用带刻度的木尺或细木条等插至罐底，经 10s 后取出，通过测量结霜的长度来估算。

4. 液氮罐的放置

液氮罐需放置在阴凉、通风且干燥的室内，不得暴晒，不可横倒放置。

5. 做好定期的清洗和保养工作

每年应清洗一次罐内杂物，将空罐放置 2d 后，用 40~50℃ 中性洗涤剂擦洗，再用清水多遍冲洗，干燥后方可使用。使用过程中，如发现罐的外壁结霜，说明罐的夹层功能失去作用，要尽快转移精液。

medium-sized liquid nitrogen tanks have a coat and a belt for carrying the back.

In general, liquid nitrogen tank should have good insulation performance, durable and easy to use, so as to better meet the needs of production.

7.3　Use of liquid nitrogen tank

7.3.1　Check carefully before use

The newly purchased or long-term unused liquid nitrogen tank must be free from damage and abnormality on the outside, dry inside without foreign matters, complete neck tube and cap plug, and intact lifting cylinder. The liquid nitrogen can be used only after it is precooled for one day and all indicators are qualified.

7.3.2　Be careful when filling liquid nitrogen

For the new tank or the tank in dry state, it must be filled slowly and precooled to prevent the tank from being damaged and reduce the service life. In order to prevent liquid nitrogen from splashing, a piece of gauze can be lined in the funnel.

7.3.3　Replenish liquid nitrogen in time

When half of the liquid nitrogen is consumed, it should be replenished immediately. The residual liquid nitrogen in the tank can be estimated by weighing, or inserted into the bottom of the tank with a calibrated wooden ruler or sliver, then taken out after ten seconds, and estimated by measuring the frosting length.

7.3.4　Placement of liquid nitrogen tank

The liquid nitrogen tank should be placed in a cool, ventilated and dry room, and should not be exposed to the sun or placed horizontally or upside down.

7.3.5　Regular cleaning and maintenance

The sundries in the tank should be cleaned once a year. After the empty tank is placed for two days, it should be scrubbed with 40-50℃ neutral detergent, then washed with clean water for several times. It can be used after drying. In the process of use, if frost is found on the outer wall of the tank, it means that the vacuum of the jar is out of order and semen should be transferred as soon as possible.

6. 防冻伤

液氮是一种超低温液体,如溅到皮肤上会引起冻伤,因此在灌充和取出液氮时应注意做好自身的防护。

7.3.6 Frostbite

Liquid nitrogen is a kind of ultra-low temperature liquid. If it is splashed on the skin, it will cause frostbite similar to burn. Therefore, we should pay attention to self-protection when filling and taking out liquid nitrogen.

任务 1 精液的品质评定
Task 1 Evaluation of Semen

任务描述

有一公猪,一次采集精液 285mL,精液呈乳白色、略有腥味,精子活力为 0.78,精子密度为 2.4 亿个/mL,精子畸形率为 12.39%。试分析该公猪的精液质量是否合格?

Task Description

In a boar, 285 mL semen was collected at one time. The semen was milky white and smelt slightly fishy. Sperm viability rate was 0.78, density was 240 million/mL, and sperm abnormality rate was 12.39%. Please analyze whether the semen of this boar is qualified?

任务实施

一、精液的感官评定

(一)准备工作

将采集好的精液做好标记,迅速置于 37℃左右的温水或保温瓶中备用。

(二)检查方法

1. 采精量

采精后应立即检测采精量。将采集的精液盛放在带有刻度的集精杯或试管中,直接读出精液量。

注意:猪、马的精液需滤除胶状物质后再检测。

2. 颜色

肉眼观察装在透明容器中的精液颜色。

Task Implementation

1 Sensory evaluation of semen

1.1 Preparations

The collected semen should be labeled and immediately placed in warm water at around 37℃ or thermos flask.

1.2 Methods

1.2.1 Volume

The volume of semen should be detected immediately after semen collection. The semen should be placed in a cup or a test tube with scales to directly read the volume of semen.

Note: The semen of boar and male horse should be filtered before detecting the volume of semen.

1.2.2 Color

Observe the color of the semen contained in a transparent container.

3. 气味

用手慢慢在装有精液的容器上方扇动，并嗅闻精液的气味。

4. 云雾状

观察装在透明容器中的精液状态，主要观察液面的变化情况。

（三）结果评定

1. 采精量

采精量的多少受多种因素影响，但超出正常范围（表2-2）太多或太少，应及时查明原因。

1.2.3　Odor

Slowly wave plam of the hand over the container and smell the odor of the semen.

1.2.4　Cloudy appearance

Observe the semen state in a transparent container, mainly observe the changes of the fluid.

1.3　Result evaluation

1.3.1　Volume

The volume of semen is influenced by many factors, but if it exceeds the normal range (Table 2-2) by too much or too little, the reasons should be found out in time.

表2-2　成年公畜（禽）的采精量
Table 2-2　Semen volume of livestock (poultry)

动物种类 Animal species	一般采精量/mL General volume/mL	正常采精范围/mL Normal range/mL
奶牛 Dairy cattle	5～10	0.5～14
肉牛 Beef cattle	4～8	0.5～14
水牛 Buffalo	3～6	0.5～12
山羊 Goat	0.5～1.5	0.3～2.5
绵羊 Sheep	0.8～1.2	0.5～2.5
猪 Pig	150～300	100～500
马 Horse	40～70	30～300
鸡 Rooster	0.5～1.0	0.2～1.5

2. 颜色

正常精液一般为乳白色或灰白色。若精液颜色异常，应当废弃，并查明原因，及时治疗。

3. 气味

正常精液略带腥味，牛、羊精液除具有腥味外，另有微汗脂味。如有异常气味，可能是混有尿液、脓汁、粪渣或其他异物，应废弃。

4. 云雾状

精液的液面呈上下翻滚状态，像云雾一样，称为云雾状。云雾状越明显，说明精液密度越大，活力越高。正常未

1.3.2　Color

The color of normal semen is usually milky white or greyish white. If the semen color is abnormal, it should be discarded. We should find out the cause and treat in time.

1.3.3　Odor

Normal semen has a slightly fishy odor, the semen of bull and ram have a slightly sweaty and fatty odor. If there is abnormal odor, it could be mixed with urine, pus, feces or others, and should be discarded.

1.3.4　Cloudy appearance

The fluid of semen rolls up and down like a cloud, so it is called cloudy appearance. The more obvious cloudy appearance of semen is, the higher

稀释的牛、羊精液肉眼可观察到云雾状。

二、精子活率评定

（一）准备工作

（1）器械准备。光电显微镜，清洗干净的载玻片和盖玻片，移液枪，吸头，擦镜纸。有条件的亦可选用全自动精子分析仪。

（2）试剂准备。生理盐水。

（3）精液准备。新鲜精液。

（二）评定方法

用移液枪吸取 1 滴原精液或经生理盐水稀释的精液，滴在载玻片上，呈 45°盖好盖玻片，载玻片与盖玻片之间应充满精液，避免气泡存在，置于显微镜下观察，估测呈直线运动的精子数占总精子数的百分率。

（三）结果评定

精子活率是指精液中呈直线运动的精子数占总精子数的百分率。评定精子活率多采用"十级评分制"法，划分为 1.0、0.9、0.8、…0.2、0.1 等 10 个等级。如果视野中 100% 的精子作直线前进运动，活率评为 1.0 级；90% 者评为 0.9，80% 者评为 0.8，依此类推。各种动物的新鲜精液的精子活率一般为 0.7～0.8，否则生产上不能用于保存和输精。

（四）注意事项

（1）牛、羊和鸡的精子密度较大，可用生理盐水稀释后再检查。

（2）精子活率是评价精液品质的一个重要指标，与受精力密切相关，一般在采精后、精液处理前后及输精前都要进行检测。

density of semen is and the higher viability of sperm is. Cloudy appearance in undiluted semen of bull and ram can be seen directly.

2 Evaluation of sperm viability

2.1 Preparations

（1）Equipment. Microscope, clean glass slides and coverslips, pipetting gun, sucker, lens wipes. A fully automatic sperm analyzer can also be used if conditions permit.

（2）Reagents. Normal saline.

（3）Semen. Fresh semen.

2.2 Methods

A drop of the original semen or diluted semen is dropped on the glass slide with a pipetting gun, covered by coverslip at an angle of 45°. The space between slide and coverslip should be filled with semen to avoid bubbles. Observe under a microscope and estimate the percentage of sperm in linear motion.

2.3 Result evaluation

Sperm viability refers to the percentage of sperm in linear motion. Sperm viability is assessed by ten-grade scoring system, which is divided into 10 grades such as 1.0, 0.9, 0.8, … 0.2, 0.1. If 100% sperm moves in a straight line, sperm viability is 1.0, 90% sperm moves in a straight line is 0.9, and so on. Viability of fresh semen in various animals is generally 0.7-0.8, otherwise, it can not be used for preservation and insemination in production.

2.4 Cautions

（1）The sperm density of bull, ram and cook is high, which can be diluted with normal saline and then examined.

（2）Sperm viability is an important index for evaluating semen quality, which is closely related to fertilization ability. Generally, it should be tested after semen collection, before and after semen treatment and before insemination.

（3）温度对精子活率影响较大，要求检查温度为37～38℃，如果没有保温装置的，检查速度要快，在10s内完成。

精子活率评定
Evaluation of sperm viability

（4）精子活率评定带有一定的主观性，应观察3～5个视野，取平均值。如果采用电视显微镜，可几人同时观察，评定结果较为准确。

三、精子密度评定

精子密度又称精子浓度，是指每毫升精液中所含有的精子数。目前，常用的评定方法有估测法、血细胞计数法和精子密度仪测定法。

（一）估测法

1. 准备工作

（1）器材准备。显微镜、载玻片、盖玻片、移液枪、吸头、擦镜纸等。

（2）精液的准备。新鲜的精液。

2. 评定方法

通常结合精子活率评定进行。取1小滴精液滴于清洁的载玻片上，盖上盖玻片，使精液分散成均匀的薄层，置于显微镜下观察精子间的空隙。

3. 结果评定

根据显微镜下精子的密集程度，把精子密度大致分为密、中、稀三个等级（图2-5）。

(3) Temperature has a great influence on sperm viability. The evaluation temperature should be about 37-38℃. If there is no heat preservation device, the evaluation should be do quickly and completed in 10 seconds.

(4) The evaluation of sperm viability is subjective. We should observe 3-5 fields of view and take the average. If a TV microscope is used, several people can observe at the same time, and the evaluation results are more accurate.

3 Evaluation of sperm density

Sperm density, also known as sperm concentration, refers to the number of sperms contained per milliliter of semen. At present, the commonly used evaluation methods are estimation method, blood count method and sperm density measurement method.

3.1 Estimation method

3.1.1 Preparations

(1) Assess equipment. Microscope, glass slides, coverslips, pipetting gun, sucker, lens wipes, etc.

(2) Semen. Fresh semen.

3.1.2 Methods

This method is usually combined with sperm viability assessment. Take a drop of semen onto a cleaned glass slide, then covered with coverslip. In this process, make sure the semen disperses into a uniform thin layer and observe the gap between sperms under the microscope.

3.1.3 Result evaluation

According to the degree of sperm concentration under the microscope, the sperm density can be roughly divided into three grades: dense, medium and sparse (Figure 2-5).

由于各种动物精子密度差异很大,很难使用统一的等级标准,检查时应根据经验,对不同的动物采用不同的标准(表 2-3)。该法具有较大的主观性,误差也较大,但简便易行,在基层人工授精站常采用。

The sperm density of various animals is quite different, so it is difficult to use a unified grade standard. Different standards should be used for different animals according to experience (Table 2-3). The method has great subjectivity and large error range, but it is simple and easy, and is often used in local artificial insemination stations.

密 Dense　　　　中 Medium　　　　稀 Sparse

图 2-5　精子密度
Figure 2-5　Sperm density

表 2-3　各种动物精子密度
Table 2-3　Sperm density of various animals

动物类别 Animal species	精子数/（亿个/mL） Sperm concentration/（100 million/mL）		
	密 Dense	中 Medium	稀 Sparse
牛 Cattle	>15	10～15	<10
羊 Sheep	>25	20～25	<20
猪 Pig	>3	1～3	<1
鸡 Rooster	>40	20～40	<20

（二）血细胞计数法

1. 准备工作

(1) 器材准备。显微镜、血细胞计数板、盖玻片、移液枪、吸头、计数器、擦镜纸等。

(2) 试剂准备。3% NaCl 溶液。

(3) 精液准备。新鲜的精液。

2. 评定方法

(1) 清洗器械。将血细胞计数板及盖玻片用蒸馏水冲洗,使其自然干燥。

(2) 稀释精液。用 3% NaCl 溶液对精液进行稀释,稀释倍数以方便计数为准。牛、羊的精液一般稀释 100、

3.2 Blood count method

3.2.1 Preparations

(1) Assess equipment. Microscope, hemacytometer, coverslips, pipetting gun, sucker, lens wipes, etc.

(2) Reagents. 3% NaCl solution.

(3) Semen. Fresh semen.

3.2.2 Methods

(1) Clean the hemocytometer and coverslips with distilled water, then allow them to dry naturally.

(2) The semen is diluted with 3% NaCl solution, and the dilution factor should be made convenient for counting. The semen of cattle and sheep is diluted 100 or 200

200倍，猪的精液稀释10、20倍。

（3）找准方格。将血细胞计数板置于载物台上，盖上盖玻片，先在100倍显微镜下查看方格全貌（图2-6），再在400倍显微镜下查找其中一个中方格。

（4）镜检。将稀释的精液滴一滴于计数室上盖玻片的边缘，使精液自动渗入计算室（图2-7），静置3min，计数具有代表性的5个中方格内的精子数。一般计数四个角的中方格和中间一个中方格（或对角线上的五个中方格）。

（5）计算。1mL原精液的精子数＝5个中方格内的精子总数×5×10×1 000×稀释倍数。

factors, and the semen of pig is diluted 10 or 20 factors.

(3) Place the hemacytometer on the objective table, then cover the coverslip. First, observe the full view of the counting chamber under low magnification (100×)(Figure 2-6), then look for one of the large double-ruled squares under a high magnification (approximately 400×).

(4) Diluted semen is dropped on the edge of the coverslip, then the semen is infiltrated automatically into the counting chamber (Figure 2-7). Wait about 3 minutes before starting to count the number of sperm in five representative squares. Generally, the middle medium square and four medium squares in the corner (or five medium squares of the diagonal) are counted.

(5) The number of sperm per milliliter is equivalent to the number of sperm counted over 5 double-ruled squares×10 000 ×the dilution factor.

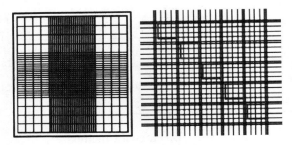

图2-6 计数室结构

Figure 2-6 Structure of counting chamber

血细胞计数法

Blood count method

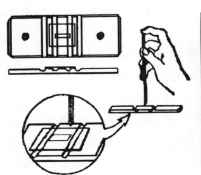

图2-7 滴加精液

Figure 2-7 Dripping semen

3. 结果评定

正常情况下，鸡的精子密度较大，为 20 亿~40 亿个/mL；羊为 20 亿~30 亿个/mL；牛为 10 亿~15 亿个/mL；猪为 1 亿~3 亿个/mL。该方法因检测速度慢，在生产上用得较少，但结果准确，一般用于结果的校准及产品质量的检测。

4. 注意事项

（1）滴入精液时，不要使精液溢出盖玻片，也不可因精液不足而导致计数室内有气泡或干燥之处。

（2）计数时，以头部压线为准，按照"数头不数尾、数上不数下、数左不数右"的原则，避免重复或漏掉（图 2-8）。

（3）为了减少误差，应连续检查两次，取其平均值，若两次计数误差大于 10%，则应做第三次检查。

3.2.3 Result evaluation

The sperm density of rooster is relatively large, which is 2 billion to 4 billion per milliliter. It is 2 billion to 3 billionin in ram, 1 billion to 1.5 billion in bull, 100 million to 300 million in boar. Because of its slow detection speed, this method is seldom used in the process of production, but the result is accurate, and is generally used for the calibration of result and the detection of product quality.

3.2.4 Cautions

（1）When dropping semen, do not let the semen overflow the coverslip, nor cause bubbles or dry areas in counting chamber due to insufficient semen.

（2）To avoid counting twice any sperm that are at the boundary of adjacent double-ruled squares, routinely count only those boundary sperm head on or within the top and left edges of each double-ruled squares (Figure 2-8).

（3）In order to reduce errors, two consecutive checks should be made and take their average values. If the two counting errors are greater than 10%, a third check should be made.

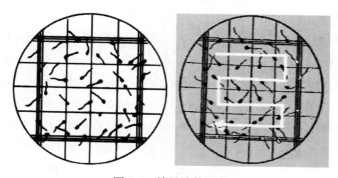

图 2-8 精子计数顺序

Figure 2-8 The order of sperm counting

（三）精子密度仪测定（图 2-9）

将待检精液样品按一定比例稀释，置于精子密度仪中读取结果。此法快速、准确、操作简便，广泛用于畜禽的精子密度测定。

3.3 Sperm density measure (Figure 2-9)

The semen samples are diluted in a certain proportion and placed in the sperm densitometer to read the results. This method is rapid, accurate and easy to operate. It is widely used for the determination of sperm density in livestock and poultry.

 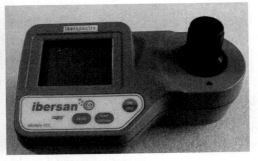

图 2-9　精子密度仪测定　　　　　　　精子密度仪测定
Figure 2-9　Sperm density measure　　　Sperm density measure

四、精子畸形率评定

（一）准备工作

（1）器械准备。显微镜、载玻片、移液枪、吸头、计数器、染色缸、擦镜纸。

（2）试剂准备。蓝墨水（或红墨水或 0.5%龙胆紫溶液）、96%酒精等。

（3）精液准备。新鲜的精液或冷冻保存的精液。

（二）评定方法

（1）抹片。用移液枪吸取精液 1 滴，滴于载玻片一端，以另一载玻片的顶端呈 35°角抵于精液滴上，精液呈条状分布在两个载玻片接触边缘之间，自右向左移动，将精液均匀涂抹于载玻片上（图 2-10）。

（2）干燥。抹片于空气中自然干燥。

（3）固定。置于 96%酒精固定液中固定 5～6min，取出冲洗后阴干。

（4）染色。用蓝（红）墨水染色 3～5min，用缓慢水流冲洗干净并使之干燥。

（5）镜检。将制好的抹片置于 400 倍显微镜下，查数不同视野的 300～

4　Evaluation of sperm abnormality

4.1　Preparations

(1) Assess equipment. Microscope, clean glass slides, pipetting gun, sucker, hand tally counter, dyeing tanks, lens wipes.

(2) Reagents. Blue ink (or red ink or 0.5% gentian violet solution), 96% alcohol, etc.

(3) Semen. Fresh semen or frozen semen.

4.2　Methods

(1) Smear. A drop of semen is dropped on one end of a glass slide with a pipetting gun. The top of the other glass slide is contacted to the semen drop at a 35° angle. The semen drop is distributed in a strip between the contact edges of the two glass slides. Then move the top glass slide from right to left and spread the semen evenly on the glass slide below it (Figure 2-10).

(2) Drying. The smear dries naturally in the air.

(3) Fixation. Fixed in 96% alcohol for 5-6 minutes. Take it out and rinse, then dry it in the shade.

(4) Staining. Dye with blue (red) ink for 3-5 minutes, rinse with slow streaming water then dry it.

(5) Microscopic examination. The smear is observed under microscope (400×), and 300-500 sperms

500个精子，记录畸形精子的数量，并计算精子畸形率。

are counted from different views. Then record the number of abnormal sperms and calculate the rate of abnormality.

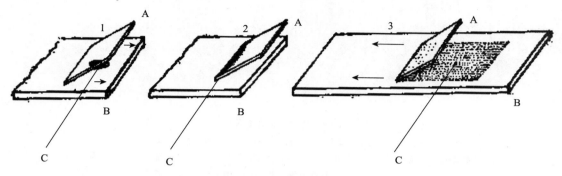

图 2-10 精液抹片示意

Figure 2-10 Semen smear

1. A 片后退，使其边缘接触精液小滴（C） 2. 精液均匀分散在 A 片的边缘

3. A 片向前推进，使精液均匀涂抹在 B 片上

1. Glass slide A is contacted to the semen drop (C)

2. The semen drop is distributed in a strip between the contact edges of glass slide A

3. Move the glass slide A from right to left and spread the semen evenly on the glass slide B

$$\text{精子畸形率（Abnormality rate）} = \frac{\text{畸形精子数 (No. of abnormal sperms)}}{\text{精子总数 (No. of sperms recorded)}} \times 100\%$$

精子畸形率评定

Evaluation of sperm abnormality

（三）结果评定

凡形态和结构不正常的精子统称为畸形精子。畸形精子类型很多，按其形态结构可分为三类（图 2-11）：头部畸形，如头部巨大、瘦小、细长、缺损、双头等；颈部畸形，如颈部膨大、纤细、曲折、双颈等；尾部畸形，如尾部膨大、纤细、弯曲、曲折、回旋、双尾等。

正常情况下，精液中会含有一定比例的畸形精子，一般牛、猪不超过 18%，羊不超过 14%，马不超过 12%。

4.3 Result evaluation

Sperms with abnormal morphology and structure are called abnormal sperm. There are many types of abnormal sperm, which can be basically divided into three types according to their morphology and structure (Figure 2-11): head abnormalities, such as large, small, slender, broken, and double-head; neck abnormalities, such as intumescent, slender, tortuous, and double-neck; tail abnormalities, such as intumescent, slender, tortuous, twisted, circling, double-tail.

Usually, there is a certain proportion of abnormal sperm in semen, it is no more than 18% in bull and boar, no more than 14% in ram and no more than 12% in male horse.

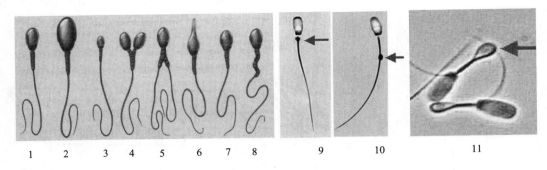

图 2-11　畸形精子类型

Figure 2-11　Types of abnormal sperm

1. 正常精子　2. 头部膨大　3. 头部瘦小　4. 双头精子　5. 双颈　6. 头部细长　7. 头部破损
8. 颈部曲折　9. 近端原生质滴　10. 远端原生质滴　11. 别针尾

1. normal sperm　2. large head　3. small head　4. double-head　5. double-neck　6. slender head　7. broken head
8. tortuous neck　9. proximal protoplasmic drop　10. distal protoplasmic drop　11. pin-shaped tail

任务 2　精液的稀释
Task 2　Dilution of Semen

任务描述

精液稀释是人工授精中的一个重要技术环节，只有经过稀释的精液，才适于保存、运输及输精。精液的保存方法不同，稀释的倍数及稀释液的成分也不相同。如何正确稀释精液？

Task Description

Semen dilution is an important technical link in artificial insemination. Only diluted semen is suitable for preservation, transportation and insemination. The methods of semen preservation are different, the dilution factors and the composition of the dilution are also different. How to dilute semen correctly?

任务实施

一、确定稀释倍数

精液稀释前需要确定稀释倍数，稀释倍数主要根据精子密度、稀释液的种类以及保存时间的长短来定。一般牛精液的稀释倍数为 10～40 倍，猪精液为 2～4 倍。

Task Implementation

1　Determine the dilution factor

Before semen dilution, it is necessary to determine the dilution factor, which is mainly determined by the sperm density, the type of dilution solution and the length of storage time. The dilution factor of bovine semen is 10-40 times, and that of boar semen is 2-4 times.

二、配制稀释液

1. 准备工作

将玻璃器皿清洗干净并晾干,然后用锡纸将瓶口封好置于120℃恒温干燥箱干燥1h,冷却备用。

2. 配制稀释液

按照配方准确称量药品,放置于烧杯中,加入蒸馏水搅拌使其充分溶解,过滤溶液并置于100℃水浴锅消毒10~20min,冷却后加入卵黄、抗生素、激素、维生素等成分,混合均匀。

三、正确稀释精液

采精后应立即对精液进行稀释,将稀释液和精液置于30℃左右环境进行同温处理。稀释时,将稀释液沿着容器壁缓缓加入精液中,边加入边摇晃,要避免剧烈震荡。稀释完毕后,应该立即进行精子活率检查。如果精子活率没有变化,则可进行分装、保存。如果精子活率下降,说明稀释液配制不当或稀释处理不当,则应弃掉精液并查明原因。

四、注意事项

(1) 配制稀释液的器具必须彻底清洗,严格消毒。

(2) 选用新鲜的蒸馏水。

(3) 药品试剂要纯净。

(4) 乳和乳粉需新鲜。将一定量的鲜乳或充分溶解的乳粉溶液,用脱脂棉进行过滤,再用水浴锅加热到92~95℃消毒灭菌10min,取出降温,除去奶皮备用。

项目二 精液的处理

Project Ⅱ Semen Treatment

2 Preparation of diluent

2.1 Preparations

The glassware should be cleaned and dried, then the bottle mouth should be sealed with tin foil and placed in a constant temperature drying oven at 120℃ for one hour, and then cooled for use.

2.2 Prepare the diluent

Drugs should be accurately weighed in beaker according to the formula, stirred with distilled water to dissolve fully, then the filtered solution should be disinfected in a water bath at 100℃ for 10-20 minutes. After cooling, egg yolk, antibiotics, hormones, vitamins and other ingredients are added and mixed well.

3 Correct dilution of semen

After semen collection, the semen should be diluted immediately, and the diluent and semen are treated at the same temperature at about 30℃. During dilution, the diluent is slowly added to semen along the wall of the container and gently shake it while adding, severe shocks should be avoid. Sperm viability should be examined immediately after dilution. If the sperm viability does not change, it can be packaged and preserved. If the sperm viability decreases, which indicates that the dilution is improperly prepared or diluted, the semen should be discarded and the cause should be identified.

4 Cautions

(1) Appliances for preparing diluents must be thoroughly cleaned and strictly disinfected.

(2) To choose fresh distilled water.

(3) Drugs should be purified.

(4) Milk and milk powder must be fresh. A certain amount of fresh milk or fully dissolved milk powder solution is filtered with absorbent cotton, and then heated to 92-95℃ for 10 minutes in a water bath. Before using it, the temperature should be cooled down and the milk skin should be removed.

（5）鸡蛋要新鲜，卵黄中不应混入蛋白和卵黄膜。经加热消毒过的稀释液，待温度降至40℃以下再加入卵黄，并注意充分溶解。

（6）抗生素、酶类、维生素、激素等需在稀释液加热灭菌后，温度降至40℃以下再加入。

（7）稀释液最好现用现配。

(5) Eggs should be fresh and egg yolk should not be mixed with protein and yolk membrane. Before adding the egg yolk, the temperature should be dropped below 40℃ after heating and disinfection. The egg yolk must be dissolved sufficiently after adding.

(6) After the diluent is heated and sterilized, and the temperature drops below 40℃, antibiotics, enzymes, vitamins, hormones can be added.

(7) After the diluent is prepared, it is better to be used soon.

任务3　冷冻精液的制作
Task 3　Production of Frozen Semen

任务描述 / Task Description

冷冻精液是利用液氮作为冷源，将精液经过特殊处理后浸泡在液氮中，可以进行长期保存和远距离运输。冷冻精液的剂型主要有颗粒冻精和细管冻精两种，后者不易污染、便于标记、适于机械化生产，解冻和输精也比较方便，生产上普遍使用。牛的细管冻精多用0.25mL剂型。大家熟悉牛冷冻精液的制作流程吗？

Frozen semen is to immerse specially treated semen in liquid nitrogen for long-term preservation and long-distance transportation. The types of frozen semen mainly include granular type and straw type. The latter is not easy to contaminate, easy to label, convenient for mechanized production, thawing and insemination, and is widely used in production. The bovine frozen semen mostly uses the size of 0.25 mL straws. Are you familiar with the production process of frozen semen?

任务实施 / Task Implementation

一、采精及精液品质检查
1　Semen collection, assessment and dilution

用假阴道采集公牛的精液，立即进行品质检查，一般要求精子活率不低于0.7，精子密度不少于8亿个/mL，精子畸形率不超过15%。

After the semen of the bull is collected with the artificial vagina, quality assessment should be carried out immediately. Generally, the sperm viability is not less than 0.7, the sperm density is not less than 800 million/mL, and the sperm abnormality is not more than 15%.

二、精液稀释

经品质检查后，及时对精液进行稀释。精液稀释采取缓慢、少量添加的方法，切忌快速、多量稀释和剧烈震荡。

1. 一次稀释法

将配制好的含有卵黄、甘油的稀释液按一定比例加入精液中，适用于低倍稀释。

2. 两次稀释法

为了缩短甘油与精子的接触时间，常采用两次稀释法。首先用不含甘油的稀释液Ⅰ和精液置于30℃环境进行同温处理，按稀释倍数进行半倍稀释；然后把稀释精液连同稀释液Ⅱ一起缓慢降温至0～5℃，在此温度下进行第二次稀释。

三、精液分装

已经稀释的精液，通过计算机控制在一体机上进行印刷、灌装及封口。细管冻精的剂型有0.25mL、0.5mL和1.0mL，国际多采用0.25mL剂型（图2-12）。管壁外打印上产地、公牛品种、公牛号、生产日期等信息。

2 Semen dilution

After quality assessment, the semen should be diluted in time. Sperm dilution should be done slowly and in a small amount. Don't do it hastily or dilute in a large amount and shock violently.

2.1 One-step dilution method

The prepared dilution containing egg yolk and glycerin is added to the semen in a certain ratio, which is suitable for low-dilution.

2.2 Two-step dilution method

In order to shorten the contact time of glycerin with sperm, two-step dilution method is often used. First, the glycerol-free diluent Ⅰ and semen are placed in a 30℃ environment for isothermal treatment, half-diluted according to the dilution factor. Then the diluted semen is slowly cooled to 0-5℃ together with the diluent Ⅱ, and the second dilution is made at this temperature.

3 Semen packaging

The diluted semen is printed, filled and sealed on the integrated machine controlled by a computer. There are several sizes of straws such as 0.25 mL, 0.5 mL, and 1.0 mL, and the 0.25 mL straw is used more frequently in the world (Figure 2-12). Outside of straw wall information such as origin, breed, bull number, production date are printed.

图2-12 细管冻精结构示意

Figure 2-12 The structure of frozen semen straws

1. 超声波封口 2. 空气柱 3. 精液柱 4. 聚乙烯醇粉 5. 棉塞

1. ultrasonic sealing 2. air column 3. semen column 4. polyvinyl alcohol powder 5. tampon

四、降温平衡

将分装好的细管精液置于0～5℃环境中平衡2.5～4h，使甘油充分渗入

4 Cooling and balancing

The tubular semen is balanced for 2.5-4 hours at 0-5℃, so that glycerol could penetrate into the sperm

精子内部，起到抗冻保护作用。

五、冻结

将平衡后的细管精液平铺在冷冻板上，距液氮面 1~2cm 处熏蒸 5~10min，然后将合格的细管冻精移入液氮罐内保存。也可利用冷冻仪制作细管冻精。

六、保存

目前，生产中普遍采用液氮为冷源，液氮罐作为贮存冻精的容器。根据液氮的挥发情况，要定期往液氮罐中添加液氮。

七、解冻及品质评定

烧杯中盛满 38℃ 左右温水，打开液氮罐，把镊子放至罐口预冷，提起提筒，迅速夹取一支冻精，放入烧杯中，并轻轻搅拌 10s 左右（图 2-13）。待冻精融化后剪去细管的封口端，装入细管输精枪中。细管冻精品质检查可按批抽样评定，精子活率应不低于 0.35。

and plays a role of anti-freezing.

5 Freezing

The balanced tubular semen is laid flat on the freezing plate and fumigated for 5-10 minutes at 1-2 cm above the liquid nitrogen surface. Then the qualified tubular frozen semen could be transferred to the liquid nitrogen container for preservation. The freezer can also be used to freeze semen.

6 Storage

At present, liquid nitrogen is widely used as cold source in production, and liquid nitrogen container is used for storing frozen semen. According to the volatilization, liquid nitrogen should be added to the container periodically.

7 Thawing and quality assessment

Fill the beaker with warm water at about 38℃, open the liquid nitrogen container, put the tweezers upon the container to pre-cool, lift the lifting barrel, quickly grab a frozen tubule, put it into the beaker, and gently stir for about 10s (Figure 2-13). After the frozen semen is dissolved, the sealing end of the tubule should be cut and loaded into an insemination gun. The quality of frozen semen can be assessed by batch sampling, and the viability should not be less than 0.35.

图 2-13 细管冻精的解冻
Figure 2-13 Thawing of frozen semen from tubules

牛冷冻精液制作流程
Production of bull frozen semen

任务 4　精液的保存与运输
Task 4　Preservation and Transportation of Semen

任务描述
Task Description

精液保存的目的是为了延长精子在体外的存活时间，便于长途运输，从而扩大精液的使用范围。精液保存的方法，按保存温度不同可分为常温保存、低温保存和冷冻保存三种。对于远距离购买精液的猪场，运输过程至关重要。将精液从甲地运往乙地，应如何运输？

The purpose of semen preservation is to prolong the survival time of sperm in vitro, facilitate long-distance transport, and expand the scope of semen use. Semen preservation methods can be divided into room temperature preservation, low-temperature preservation and cryopreservation. Transportation is critical for pig farms that purchase semen over long distances. How should semen be transported from place A to place B?

任务实施
Task Implementation

一、猪精液的保存与运输
1　Preservation and transportation of boar semen

（一）精液的分装
1.1　Semen packaging

精液稀释后按每头母猪一次的输精量进行灌装。精液分装常采用瓶装和袋装两种。瓶装的精液分装简单方便，易于操作，容量最高为100mL（图2-14）；袋装的精液分装一般需要专门的精液分装机，用机械分装、封口，容量一般为80mL（图2-15）。

After the semen is diluted, it is packaged according to the amount of insemination in each sow. Semen packaging is often used both bottles and bags. Bottled semen is easy to install and handle. The bottled semen volume is up to 100 mL (Figure 2-14). The bagged semen is packed generally with a special semen packer, which is mechanically packed and sealed. The semen volume is typically 80 mL (Figure 2-15).

分装后的精液，要逐个粘贴标签。为了便于区分，一般一个品种一个颜色。分好后将精液瓶加盖密封，封口时尽量挤出瓶中空气，贴上标签，标明公猪的品种、耳号、采精日期、保存有效期、生产单位及相应的使用说明。

After packaging semen, the label should be affixed one by one, in order to distinguish easily, generally one breed has one color. When sealing, the air in the bottle should be squeezed out, and the label should have the information of boar's breed, ear number, date of collection, expiration date, manufacturer and corresponding instructions.

图 2-14　分装后的瓶装精液
Figure 2-14　Bottled semen

猪精液分装
Boar semen packaging

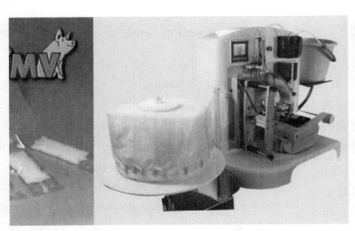

图 2-15　猪精液自动灌装打印一体机
Figure 2-15　Semen automatic filling and printing machine

（二）精液的保存

分装好的精液先置于 22~25℃ 的环境下放置 1~2h，然后移入 16~18℃ 的恒温箱中保存（图 2-16）。存放时，不论是瓶装的还是袋装的，均应平放，目的是为了增大精子沉淀后铺开的面积，减少沉淀的厚度，降低精子死亡率。

1.2　Semen preservation

The packaged semen should be placed at 22-25℃ for 1-2 hours, and then stored in a thermostat at 16-18℃ (Figure 2-16). When storing, whether it is bottled or bagged, it should be laid flat in order to increase the area of sperm precipitation, and reduce the thickness of sperm and the rate of sperm death.

项目二 精液的处理
Project Ⅱ Semen Treatment

图 2-16 猪精液保存恒温箱
Figure 2-16 Thermostat for boar semen preservation

在保存期内要注意三点：一是尽量减少恒温箱门的开关次数，防止频繁改变温度对精子的打击；二是每隔 12h 轻轻翻动一次，以防止精子沉淀；三是每天检查恒温箱内温度计的变化，防止温度出现明显的波动。若保存过程中出现停电现象，应全面检查贮存的精液品质。精液保存时间的长短，因稀释液成分的不同而异，一般可保存 2～3d。

（三）精液的运输

精液运输是猪场人工授精顺利进行的必要环节，具有提高种公猪利用率、防止疾病传播、更新猪群血液等优点。猪精液运输需注意以下几点：

（1）运输的精液需按规定进行稀释和保存，有详细的说明书，标明站名、公猪品种和编号、采精日期、精液剂量、稀释倍数、精子活率和密度等。

（2）精液包装要严密，不能发生泄

Pay attention to three points during the storage period. First, reduce the number of opening and closing of the thermostat door to prevent frequent changes in temperature from hitting the sperm. Second, gently flip once every 12h to prevent sperm precipitation. Third, check the changes of thermometer in thermostat every day to prevent significant fluctuations in temperature. If there is a power outage during storage, semen quality should be checked comprehensively. The duration of semen preservation varies with the composition of the diluent. Generally, semen can be stored for 2-3 days.

1.3 Semen transportation

Semen transportation is an essential link for the successful artificial insemination in pig farms. It has the advantages of improving the utilization rate of boars, preventing the spread of diseases and refreshing the group blood. Pay attention to the following points when transporting boar semen.

(1) The transported semen should be diluted and preserved according to regulations. Detailed specifications include the station name, boar breed and number, date of semen collection, semen volume, dilution factor, sperm viability and density.

(2) Semen packaging should be tight to avoid

漏，运输多采用精液运输箱（图 2-17）或广口保温瓶。

（3）尽量避免在运输过程中产生剧烈震荡和碰撞。

（4）精液运输过程中要注意保持温度的恒定。

（5）精液运输过程中要防止阳光直射到泡沫箱，更不能直射到输精瓶上。

leakage, and the semen transport box (Figure 2-17) or wide-mouth thermos bottle is used for transportation.

(3) Try to avoid violent shocks and collisions during transportation.

(4) Pay attention to keeping the temperature constant during the transportation.

(5) During the transportation, it is necessary to prevent direct sunlight from reaching the foam box, and especially the bottle.

图 2-17　猪精液运输箱

Figure 2-17　Transport box of boar semen

二、牛精液的保存与运输

（一）精液的保存

生产中多采用液氮罐贮存冻精。该法保存时间长，精液的使用不受时间、地域以及种公牛寿命的限制，对人工授精技术的推广及现代畜牧业的发展均具有十分重要的意义。

取用冻精时，操作要敏捷迅速，将镊子在液氮罐口预冷，冻精不可提出液氮罐口，冻精脱离液氮的时间不得超过20s，取完后要迅速将余下的冻精再次浸入液氮内，及时盖上罐塞。在向另一个液氮贮存罐内转移冷冻精液时，冻精

2　Preservation and transport of bull semen

2.1　Semen preservation

Liquid nitrogen container is often used for storing frozen semenin. This method has a long preservation time, and the use of semen is not limited by time, region and life of the bull. It is of great significance for the promotion of artificial insemination and the development of modern animal husbandry.

When taking frozen semen, the operation should be quick and agile. Tweezers should be pre-cooled in the mouth of liquid nitrogen container. The frozen semen should not be put out of the liquid nitrogen container. The time for the frozen semen to leave the liquid nitrogen should not exceed 20 seconds. After taking the frozen semen, the rest of them should be im-

在空气中转移时间不得超过 5s，并迅速将精液再次浸入液氮内。

（二）精液的运输

由冷冻精液站提供的冷冻精液必须经过精子活率抽检，查验种公牛品种、公牛号及数量，与标签一致无误后方可运输。盛装精液的液氮罐必须确保其保温性能，加外保护套或装入液氮罐运输箱（图 2-18）内，罐与罐之间要用填充物隔开，在罐底加防震软垫，防止颠簸撞击，严防倾倒，装卸车时要严防液氮罐碰击，切不能在地上随意拖拉，以免损坏液氮罐，降低使用寿命。运输途中应随时检查并及时补充冷源。

mersed in the liquid nitrogen again quickly, and the mouth should be covered in time. When transferring frozen semen to another liquid nitrogen container, the frozen semen should not stay in the air for more than 5 seconds, and the semen should be quickly immersed in liquid nitrogen.

2.2 Semen transportation

The frozen semen provided by the frozen semen station must be sampled for sperm viability, and the bull breed shall be checked, along with the bull number and number of frozen semen, and it can be transported only if it is in accordance with the label. The liquid nitrogen container must ensure its thermal insulation performance, outer protective jacket should be added or be packed in the liquid nitrogen container transport box (Figure 2-18). The containers should be separated by fillers, and the bottom of the container should be shockproof. Prevent bumps and impacts, avoid tumbling, prevent the liquid nitrogen container from being hit, never drag on the ground when loading and unloading, so as not to damage the liquid nitrogen container and reduce the service life. The liquid nitrogen should be inspected and replenished at any time during transportation.

图 2-18 液氮罐运输箱

Figure 2-18 Transport box of liquid nitrogen container

项目三　发情鉴定与输精
Project Ⅲ　Estrus Identification and Insemination

项目导学

发情鉴定是动物繁殖技术中最为基础和关键的技术。通过掌握雌性动物的生殖器官构造和机能、卵泡的发育规律、雌性动物的发情生理及生殖激素对发情的调节等相关知识，做好发情鉴定，确定最佳的配种时间，规范完成输精工作，提高受配率和受胎率。

Project Guidance

Estrus identification is the most basic and key technology in animal reproduction. Learn the anatomy and function of the female reproduction organs, the growth of the follicles, estrous cycle, and the regulation of reproductive hormones, etc., so as to detect the time of estrus, determine proper time for insemination, standardize the insemination procedure, improve the rate of fertilization and conception.

学习目标

>>> 知识目标

- 理解雌性动物生殖器官的组成和主要生理功能。
- 掌握生殖激素的概念、分类及应用。
- 掌握雌性动物的发情生理和规律。
- 了解卵泡的发育和排卵机理。
- 理解生殖激素对发情周期的调节机理。

>>> 技能目标

- 能准确进行牛、羊和猪的发情鉴定。
- 规范完成对输精器械的清洗和消毒。
- 规范完成各畜禽的输精操作。

Learning Objectives

>>> Knowledge Objectives

- To understand the anatomy and main function of female reproduction organs.
- To master the concept, classification and application of reproduction hormones.
- To master the physiology regulation of estrus.
- To know the growth of the follicles and the ovulation time.
- To understand the regulatory effects of reproductive hormones on estrus cycle.

>>> Skill Objectives

- To master the technology of estrus detection in cows, ewes and sows.
- To standardize the cleaning and disinfection of insemination devices.
- To standardize the insemination of livestock and poultry.

一、雌性动物生殖器官的结构与功能

雌性动物的生殖器官主要由卵巢、输卵管、子宫、阴道、阴唇、阴蒂等组成。几种常见母畜生殖器官见图3-1。母禽的生殖器官仅包括卵巢和输卵管两部分，家禽只有左侧生殖器官，右侧生殖器官在孵化的第7～9天就停止发育，到孵出时已退化，仅留残迹。

1 Structure and functions of female reproductive organs

The female reproductive organs consist of the ovary, oviduct, uterus, vagina, vulva and clitoris (Figure 3-1). The hen has a pair of ovaries and oviducts. On the 7th to 9th day of hatching, the right-side ovary and oviduct will stop developing until only a vestige remains in the end. Therefore, the sexually mature hen only has a well-developed ovary and oviduct on the left side.

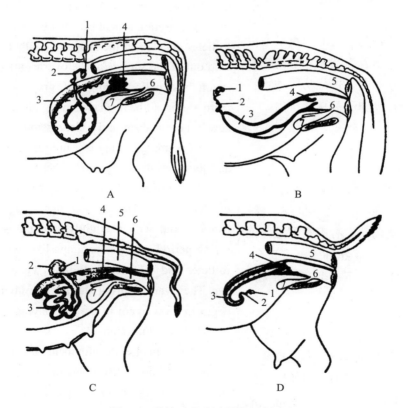

图 3-1 雌性动物生殖器官解剖示意

Figure 3-1 Reproductive organs of the female farm mammal

A. 牛 cow B. 马 mare C. 猪 sow D. 羊 ewe

1. 卵巢 2. 输卵管 3. 子宫角 4. 子宫颈 5. 直肠 6. 阴道 7. 膀胱

1. ovary 2. oviduct 3. uterus horn 4. cervix 5. rectum 6. vagina 7. urinary bladder

（一）卵巢

1. 形态和位置

卵巢是雌性动物产生卵子和分泌激素的重要结构，左右各一，附着在卵巢系膜上。各种母畜的卵巢形状如图3-2所示。

牛的卵巢为扁椭圆形，位于子宫角尖端的两侧。羊的卵巢比牛的圆而小，位置与牛相同。初生仔猪的卵巢呈肾形，淡红色，位于荐骨岬两旁稍后方或在骨盆腔前口两侧的上部；当接近性成熟时（4～5月龄），因卵泡发育而呈桑葚形，性成熟后卵泡突出于卵巢表面，此时卵巢似串状葡萄。马的卵巢呈蚕豆形，表面光滑，附着缘宽大，游离缘有一凹陷，称为排卵窝，这是马属动物特有的一个结构，其卵子只从此排出。

禽卵巢以短的系膜附着于左肾前部。幼禽为扁平形，表面略呈颗粒状。产蛋期卵巢如葡萄状，有肉眼可见卵泡1 000～1 500个，成熟卵泡4～5个。停产期卵巢回缩，到下一个产蛋期又开始生长。

2. 机能

（1）卵泡发育和排卵。卵巢皮质部表层分布着许多原始卵泡，原始卵泡经过初级卵泡、次级卵泡、三级卵泡、成熟卵泡等几个发育阶段，部分卵泡最终发育为成熟卵子，由卵巢排出，并在原卵泡腔处形成黄体。

（2）分泌雌激素和孕激素。卵泡发

1.1 Ovary

1.1.1 Morphology and location

Ovaries produce ovum and secrete hormones. It adheres to the mesovarium of ovary both with one on the left side and the other on the right. Different breeds of females may have different shapes of ovaries (Figure 3-2).

The ovaries of cow mainly shape like flat ellipse, located on both sides of the tip of uterine horns. The ovaries of ewe are smaller to those of cow, and they are in the same position as cow. The ovaries of piglets are kidney-shaped and reddish in color. They are located at the posterior sides of the sacrum or at the upper sides of the anterior opening of the pelvic cavity. When becoming mature (about 4-5 months old), the follicles develop in the shape of mulberries, which extend out of the surface of the ovaries after sexual maturation, resembling strings of grapes. The ovaries of mare are broad bean-shaped with smooth surface, wide attach-ment margin and a depression (ovulation fossa) in the free margin. This is a unique structure of mare, and its ovum discharged from here.

The ovaries of poultry attach to the anterior part of the left kidney with a short mesangium. The poultry's ovary is flat with a slightly granular surface. During laying period, ovaries are grape-like, with 1 000-1 500 follicles and 4-5 mature follicles visible to the naked eye. The ovaries shrink during non-laying period and begin to grow again in the next laying period.

1.1.2 Function

(1) Follicular development and ovulation. There are many primordial follicles in the surface of ovarian cortex. It passes through several stages of development, such as primary follicle, secondary follicle, tertiary follicle and mature follicle. Some follicles eventually develop into mature eggs. After the ovum is discharged from the mature follicle, and the cavity of follicle changes into corpus luteum.

(2) Secretion of estrogen and progesterone. During

育过程中，卵泡内膜细胞可分泌雌激素，雌激素是导致母畜发情的直接因素。黄体可分泌孕激素，孕激素是维持母畜妊娠所必需的激素之一。

follicular development, follicular endometrial cells can secrete estrogen, which is the direct factor leading to estrus in female animals. The corpus luteum can secrete progesterone, which is one of the hormones to maintain the pregnancy.

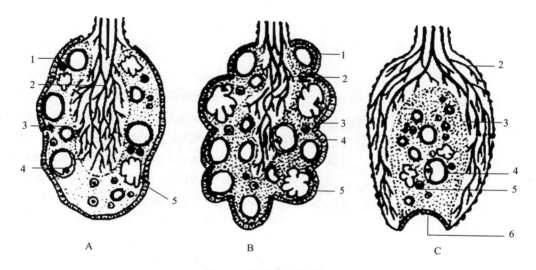

图 3-2 母畜的卵巢形状

Figure 3-2　Different shapes of female livestock's ovaries

A. 牛 cow　B. 猪 sow　C. 马 mare

1. 生殖上皮　2. 髓质部　3. 皮质部　4. 卵泡　5. 黄体　6. 生殖上皮（排卵窝）

1. reproductive epithelium　2. medulla　3. cortical　4. follicle　5. corpus luteum

6. reproductive epithelium（ovulation fossa）

（二）输卵管

1. 形态和位置

输卵管位于卵巢和子宫角之间，形状弯曲，长 15～30cm。输卵管可分为漏斗部、壶腹部和峡部 3 个部分。

2. 机能

由卵巢排出的卵子先被输卵管伞部接纳（图 3-3），然后向壶腹部运送。同时进入输卵管的精子获得受精能力，在壶腹部完成受精作用，并进行早期卵裂。

1.2　Oviduct

1.2.1　Morphology and location

The oviduct is located between the ovary and the uterine horn, with a curved shape and a length of 15-30 cm. It consists of infundibulum, ampulla and isthmus.

1.2.2　Function

The ovum discharged from the ovaries is caught by the umbrella of oviduct and then transported to ampulla (Figure 3-3). At the same time, the sperm entering the oviduct acquires fertilization ability and completes fertilization, then the early cleavage begins.

图 3-3 输卵管伞部接纳卵子
Figure 3-3　Ovum caught by the umbrella of oviduct

（三）子宫

1. 形态和位置

子宫大部分位于腹腔，小部分位于骨盆腔，背侧为直肠，腹侧为膀胱。多数动物的子宫都由子宫角、子宫体和子宫颈三部分组成。牛、羊的子宫角弯曲如绵羊角（图 3-4A、图 3-4B），两角基部之间有纵隔将两子宫角分开，称为对分子宫（也称双间子宫）。猪的子宫有两个长而弯曲的子宫角（图 3-4C、图 3-4D），两角基部之间的纵隔不明显，为双角子宫。

2. 机能

子宫可以筛选、贮存和运送精子，促进精子获能；有利于孕体的附植、妊娠和分娩；调节卵巢黄体功能，引起发情等。

1.3　Uterus

1.3.1　Morphology and location

Uterus is mostly located in the abdominal cavity, with a small part in the pelvic cavity. The dorsal side is rectum and the ventral side is bladder. The uterus of most animals consists of uterus horn, uterus body and uterus cervix. The uterus horns of cow and ewe are crooked like the horns of sheep (Figure 3-4A and Figure 3-4B). There is a mediastinum at the uterus corner base. This kind of uterus is also called bifurcated uterus. The pig's uterus has two long, curved horns. The mediastinum of sow is not obvious between the two uterus corner bases, and it is called bicornute uterus (Figure 3-4C and Figure 3-4D).

1.3.2　Function

The uterus has the following functions: ①screening, storing and transporting the sperm, promoting sperm capacitation. ②beneficial to conceptus transplants, pregnancy and delivery. ③regulating ovarian luteal function, leading to estrus and other functions.

 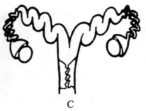 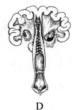

图 3-4　子宫　　　　　　　　　　　　　　牛的子宫
Figure 3-4　Uterus　　　　　　　　　　　Uterus of cow
A, B. 牛 cow　C, D. 猪 sow

（四）阴道

阴道既是交配器官，也是分娩时的产道。其前接子宫颈阴道部，后接尿生殖前庭。阴道还是子宫颈、子宫黏膜和输卵管分泌物的排出管道。在发情周期中，阴道的黏膜表面会发生变化，动物开始发情时期非常潮湿，后逐渐变得干燥，甚至黏稠。阴道内的生化和微生物环境，能保护上生殖道免受微生物的入侵。

二、生殖激素

激素是由机体产生，经体液循环或空气传播等途径作用于靶器官或靶细胞，具有调节机体生理机能的微量信息传递物质或微量生物活性物质。直接作用于生殖活动，并以调节生殖机能、生殖过程为主要功能的激素，称为生殖激素（表3-1）。

雌性动物卵子的生成、卵泡的发育、卵子的排出、发情周期的变化、受精、妊娠、分娩及泌乳等生理活动都与生殖激素的作用有密切的关系。生殖激素不仅在各项繁殖技术上广泛应用，而且可作为一种调节剂，制成各种激素类药物来调节动物的发情，防治繁殖障碍，尤其是治疗不孕症，效果非常明显。

1.4 The vagina

The vagina serves as the female organ of copulation at mating and as the birth canal at parturition. It connects with the cervix in the anterior and extends to urethral reproductive vestibule in the posterior. The vagina serves the dual role of a passageway for the reproductive and urinary systems. Its mucosal surface changes during the estrous cycle from very moist when the animal is ready for mating to almost dry, even sticky. Its biochemical and microbial environment can protect the upper genital tract from microbial invasion.

2 Reproductive hormones

Hormone is a chemical substance produced in the body that controls and regulates the activity of certain cells or organs. The hormone that directly acts on reproductive activities and regulates reproductive function and reproductive process is called reproductive hormone(Table 3-1).

Reproductive hormones are closely related to the physiological activities of female animals, such as ovum formulation, follicles development, ovulation, estrus cycle changes, fertilization, pregnancy, parturition and lactation. Reproductive hormones are not only widely used in various reproductive technologies, but also used as a kind of regulator, which is useful to regulate animal estrus, prevention and control of reproductive disorders. It shows an effective result in the treatment of infertility especially.

表 3-1 主要生殖激素的名称、来源及生理功能
Table 3-1 Name, source and function of main reproductive hormones

名称 Name	简称 Abbreviation	来源 Source	生理功能 Function
促性腺激素释放激素 Gonadotropin releasing hormone	GnRH	下丘脑 Hypothalamus	促进 LH 和 FSH 释放 Stimulate the release of LH and FSH

(续)

名称 Name	简称 Abbreviation	来源 Source	生理功能 Function
催产素 Oxytocin	OXT	下丘脑 Hypothalamus	促进子宫收缩和排乳 Promote uterine contraction and lactation
促卵泡素 Follicle-stimulating hormone	FSH	垂体前叶 Anterior pituitary	促进卵泡发育和精子发生 Promote follicular development and spermatogenesis
促黄体素 Luteinizing hormone	LH	垂体前叶 Anterior pituitary	促进排卵，形成黄体，促进孕酮分泌 Promote ovulation, corpus luteum formation and progesterone secretion
促乳素 Prolactin	PRL	垂体前叶 Anterior pituitary	促进泌乳，增强母性行为 Promote lactation, enhance maternal behavior
孕马血清促性腺激素 Pregnant mare's gonadotrophin	PMSG	马胎盘 Horse placenta	具有 FSH 和 LH 双重活性，以 FSH 为主 Double bio-activities of FSH (main) and LH
人绒毛膜促性腺激素 Human chorionic gonadotrophin	HCG	灵长类胎盘绒毛膜 Primate placental chorion	与 LH 相似 Similar to LH
雌激素 Estrogen	E_2	卵巢、胎盘 Ovary and placenta	促进动物发情行为，维持第二性征，刺激生殖器官的发育 Promote the estrus behavior, maintain the secondary sexual characteristics and stimulate the development of reproductive organs
孕激素 Progesterone	P_4	卵巢（黄体）、胎盘 Corpus luteum and placenta	调节发情，维持妊娠，促进乳腺发育 Regulate estrus, maintain pregnancy and promote mammary development
雄激素 Androgen	A	睾丸间质细胞 Interstitial cells of testis	维持雄性动物第二性征和性行为，促进精子发生 Maintain the secondary sexual characteristics and sexual behavior of male animals, promote spermatogenesis
松弛素 Relaxin	RLX	卵巢、胎盘 Ovary and placenta	促使子宫颈扩张、骨盆韧带松弛 Promote cervical dilatation and pelvic ligament relaxation
前列腺素 Prostaglandin	PG	子宫（内膜） Uterus (endometrium)	溶解黄体，促进子宫收缩 Regress corpus luteum, promote uterine contraction
外激素 Pheromone	PHE	外分泌腺 Exocrine glands	影响性行为和性活动 Influence sexual behavior

三、雌性动物的性机能发育

雌性动物的性机能发育是一个由发生、发展直至衰退停止的过程，一般分为初情期、性成熟期、体成熟期及繁殖机能停止期。为了指导生产，还涉及初配期。

1. 初情期

初情期是指雌性动物第一次出现发情现象和排卵的时期。到达初情期体重为成年体重的30%~40%，虽有发情表现，但不完全，发情周期也往往不正常，其生殖器官仍在继续生长发育中，所产生的卵子质量较差。此阶段配种虽有受精的可能性，但不宜配种。

2. 性成熟期

初情期后，随着年龄的增长，生殖器官进一步发育成熟，发情排卵活动已趋正常，具备了正常繁殖后代的能力，此时称为性成熟。性成熟期时期，动物体重占成年体重的50%~60%，身体发育尚未成熟，故一般不宜配种。

3. 体成熟期

一般当雌性动物体重达到成年体重的70%时就可以配种。雌性动物在初配后受胎，身体仍未完全发育成熟，经过一段时间后才能达到体成熟。

4. 繁殖机能停止期

雌性动物经过多年的繁殖活动，生殖器官逐渐老化，繁殖机能逐渐衰退，甚至丧失繁殖能力。在畜牧生产中，一般在母畜繁殖机能停止之前，只要生产效益明显下降，就对其进行淘汰。

3 Sexual development of female animal

Sexual development of female animal is a process including the puberty stage, sexual maturity stage, adult stage and reproductive cessation stage. In practical production, we also consider the initial mating stage.

3.1 Puberty stage

Puberty stage refers to the period when the female animal appears estrus and ovulation for the first time. The body weight is about 30%-40% of the adutt's body weight at this stage. The estrous behavior is not complete, and the estrus cycle is often abnormal. Their reproductive organs are still growing and developing, and the quality of ovum is not good. In conclusion, female is not suitable for mating at this stage.

3.2 Sexual maturity stage

After the puberty stage, as the female grows up, reproductive organs are getting mature, estrus and ovulation activities are more regular. At this stage, the female has the ability of reproduction. During sexual maturity stage, the body weight accounts for 50%-60% of the adult's body weight, which means the body has not mature yet, so it is still not suitable for mating.

3.3 Adult stage

When the body weight of female animal accounts for 70% of the adult's, we can use it for breeding. After initial mating, the body is still not fully mature. It takes some time for them to reach full body maturity.

3.4 Reproductive cessation stage

After years of reproduction, the ageing organs of female animals' reproductive system are gradually losing their reproductive ability. In animal husbandry production, we usually eliminate those female animals as long as their production efficiency obviously decreases. This usually happens before the reproductive cessation stage.

四、卵子的发生与卵泡发育

（一）卵子的发生

1. 卵原细胞的增殖

动物在胚胎期性别分化后，雌性胎儿的原始生殖细胞便分化为卵原细胞。卵原细胞通过多次有丝分裂的方式增殖成许多初级卵母细胞。

2. 卵母细胞的生长

初级卵母细胞的生长是伴随卵泡的生长而实现的，卵泡细胞为卵母细胞的生长提供营养。初级卵母细胞阶段的主要特点：卵黄颗粒增多，卵母细胞体积增大，出现透明带，卵母细胞周围的卵泡细胞通过有丝分裂增殖，由扁平形单层变为立方形多层。初情期到来之前，卵母细胞生长发育处于停滞状态。

3. 卵母细胞的成熟

卵母细胞的成熟需经过两次成熟分裂。卵泡中的卵母细胞是一个初级卵母细胞，在排卵前不久完成第一次成熟分裂。大多数动物在排卵时，卵子尚未完成成熟分裂。牛、绵羊和猪的卵子，在排卵时只是完成第一次成熟分裂，即卵泡成熟破裂时，放出次级卵母细胞和一个极体，排卵后次级卵母细胞开始第二次成熟分裂，直到精子进入透明带，卵母细胞被激活后，放出第二极体，这时才完成第二次成熟分裂。一个初级卵母细胞经分裂最终发育形成1个卵子和3个极体。

（二）卵子的形态和结构

1. 卵子的形态和大小

哺乳动物的卵子为圆球形，卵子较

4 Oogenesis and follicular development

4.1 Oogenesis

4.1.1 Oogonia proliferation

After sex differentiation in the embryonic stage, the primordial germ cells of the female fetus differentiate into oogonia. Through multiple mitosis, oogonia proliferate into many primary oocytes.

4.1.2 Oocyte growth

The growth of primary oocyte is accompanied by the growth of follicle, which provides the nutrition for the growth of oocyte. There are some main characteristics of primary oocyte stage, including: the number of yolk granules increases, the volume of oocyte increases, zona pellucida appears, and the follicular cells surrounding oocyte proliferate through mitosis, changing from flat monolayer to cubic multilayer. Before the puberty, the development of oocyte is arrested at primary oocyte stage.

4.1.3 Oocyte maturation

The process of oocyte maturation needs two maturation divisions. The oocyte in the follicle is a primary oocyte with completion of the first maturation division shortly prior to ovulation. The ovum has not completed maturation division yet while most animals ovulating. The ovum from cow, ewe and sow only has completed the first maturation division when ovulation occurs. In details, one secondary oocyte and one first polar body are released at the time of the follicle maturation and rupture. After ovulation, the secondary oocyte begins the second maturation division until the sperm enters the zona pellucida. Until the sperm enters the zona pellucida and the oocyte is activated, the second polar body is released, and then the second maturation division is completed. Eventually, one primary oocyte develops to form one ovum and three polar bodies.

4.2 Morphology and structure of ovum

4.2.1 Shape and size of ovum

The ovum of mammal is spherical with 70-140 μm in

一般细胞含有更多的细胞质，大多数哺乳动物的卵子，直径为 70～140 μm。

2. 卵子的结构

卵子的主要结构包括放射冠、透明带、卵黄膜及卵黄等部分（图 3-5）。

4.2.2　Ovum structure

The main structure of the ovum includes the corona radiate, the zona pellucida, the vitelline membrane and the ooplasm, et al. (Figure 3-5)

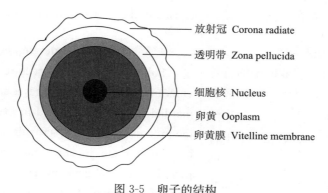

图 3-5　卵子的结构
Figure 3-5　Structure of the ovum

（1）放射冠。卵子周围致密的颗粒细胞呈放射状排列，故名放射冠。放射冠位于卵子最外层，为卵母细胞提供养分并进行物质交换。

（2）透明带。为一均质的蛋白质半透明膜，一般认为它由卵泡细胞和卵母细胞形成的细胞间质组成。其作用是保护卵子，在受精时发生透明带反应，防止多个精子进入，使受精正常进行。同时，透明带存在与精子特异性结合的位点，使异种动物的精子不能和透明带结合。

（3）卵黄膜。是卵母细胞的皮质分泌物，其作用主要是保护卵母细胞完成正常的生命活动，以及在受精过程中发生卵黄膜封闭作用，阻止多精子受精。

（4）卵黄。排卵时卵黄较接近透明带，受精后卵黄收缩，透明带和卵黄膜

(1) Corona radiate. The dense cumulus cells surround the ovum radially, which is called corona radiate. The corona radiate is located at the outermost layer of the ovum, providing nutrients to the oocyte and exchanging substances.

(2) Zona pellucida. The zona pellucida is a homogeneous translucent membrane, made of intercellular substances formed by follicular cells and oocytes generally. It protects ovum, causes the acrosome reaction during fertilization, prevents post-fertilization polyspermy. The specific binding region of sperm appears in the zona pellucid, which prevents xenogeneic species' sperm from binding with the zona pellucid.

(3) Vitelline membrane. It is the secretion of the oocyte cortex, which can protect the oocytes to complete the normal life events and undergo the vitelline membrane reaction to block polyspermy during fertilization.

(4) Ooplasm. It is closer to the zona pellucida at the time of ovulation. After fertilization, the vitelline

之间的卵黄周隙扩大，排出的极体即在周隙之中。此外，卵黄为卵子和早期胚胎发育提供营养物质。

（5）卵核。位于卵黄内，雌性动物的主要遗传物质就分布在卵核内。

（三）卵泡的生长发育

卵泡是包裹卵母细胞或卵子的特殊结构。在胚胎期，雌性动物就已形成大量原始卵泡贮存于卵巢皮质部，卵泡的生长贯穿于胚胎期、幼龄期和整个生育期。卵泡发育从形态上可分为原始卵泡、初级卵泡、次级卵泡、三级卵泡和成熟卵泡5个阶段（图3-6）。

1. 原始卵泡

原始卵泡位于卵巢皮质外周，是体积最小的卵泡，其核心为一卵母细胞，周围为一单层扁平状的卵泡上皮细胞，没有卵泡膜和卵泡腔。在胎儿期已有大量原始卵泡作为储备，除少数发育成熟外，其他均在发育过程中闭锁、退化而死亡。

2. 初级卵泡

初级卵泡由原始卵泡发育而成，排列在卵巢皮质外围，是由卵母细胞和周围的一层立方形卵泡细胞组成，卵泡膜尚未形成，也无卵泡腔。

3. 次级卵泡

在生长发育过程中，初级卵泡移向卵巢皮质的中央，卵泡上皮细胞增殖形

shrinks, and the perivitelline space between the zona pellucida and vitelline membrane becomes expanded, where the released polar bodies appear. In addition, yolk provides nutrients for ovum and early embryo development.

(5) Nucleus. The mainly genetic material is carried in the nucleus, which is located in the vitelline.

4.3 Growth and development of follicular

Follicle is a special structure around oocyte or ovum. In the embryonic stage, a large number of primordial follicles are located in the ovarian cortex, and the process of follicular growth occurs throughout the embryonic stage, juvenile stage and the child bearing period. Follicular development in morphology can be divided into five stages: primordial follicle, primary follicle, secondary follicle, tertiary follicle and mature follicle (Figure 3-6).

4.3.1 Primordial follicle

It is the smallest follicle, located in the outer periphery of the ovarian cortex. The core of primitive follicles is an oocyte surrounded by a single layer of flat follicular epithelial cells. It has no follicular membrane or follicular cavity. There are a large number of primordial follicles as reserves in the fetal period. Except for a few mature, the others are locked, degenerated and die during development.

4.3.2 Primary follicle

It develops from the primordial follicle, arranging on the periphery of the ovarian cortex, and is composed of the oocyte surrounded by a layer of cuboid follicular cells. In primary follicle, the follicular membrane has not yet formed and the follicular cavity is absent.

4.3.3 Secondary follicle

During growth and development, the primary follicle moves to the center of the ovarian cortex, and the

成多层立方形细胞，细胞体积变小，称为颗粒细胞。随着卵泡的生长，卵泡细胞分泌的液体聚集在卵黄膜与卵泡细胞（或放射冠细胞）之间形成透明带。此时尚未形成卵泡腔。

4. 三级卵泡

三级卵泡由次级卵泡发育而成。在这一时期中，卵泡细胞分泌的液体，使卵泡细胞之间分离，并使卵母细胞之间的间隙增大，形成不规则的腔隙，称为卵泡腔。随着卵泡液的增多，卵泡腔也逐渐扩大，卵母细胞被挤向一边，并被包裹在一团颗粒细胞中，形成半岛突出于卵泡腔中，称为卵丘。其余的颗粒细胞紧贴于卵泡腔的周围，形成颗粒层。

5. 成熟卵泡

三级卵泡继续生长，卵泡壁变薄，卵泡液增多，卵泡腔增大，卵泡扩展到整个卵巢的皮质部而突出于卵巢的表面。各种动物成熟卵泡的大小差异很大，牛的直径为10～14mm，猪为8～12mm，绵羊为5～10mm，山羊为7～10mm。多胎动物在一个发情周期里，可有数个至数十个原始卵泡同时发育到成熟卵泡，而单胎动物一般只有一个卵泡发育成熟并排卵。

（四）卵泡的闭锁与退化

动物出生前，卵巢上就有很多原始卵泡，但只有少数卵泡能够发育成熟并排卵，绝大多数卵泡发生闭锁和退化。越是年轻的动物，卵泡闭锁的发生越严重。卵泡发生闭锁的原因可能与垂体前

follicular epithelial cells proliferate to form a several layers of cuboidal cells with smaller volume. As the follicles grow, the fluid secreted by the follicular cells accumulates between the vitelline membrane and the follicular cells (or the corona radiate cells) to form zona pellucida. The follicular cavity has not yet formed.

4.3.4 Tertiary follicle

It develops from the secondary follicles. In this period, the fluid secreted by the follicular cells separates the follicular cells from each other, expands the space between the oocytes, and eventually forms an irregular cavity, which is called the follicular cavity. As the follicular fluid increases, the follicular cavity gradually expands. The oocyte squeezed to one side is surrounded by a mass of granulosa cells, forming a peninsula into the follicular cavity, which is called the cumulus. The other granulosa cells are closely attached to the periphery of the follicular cavity to form the granular layer.

4.3.5 Mature follicle

The tertiary follicles continue to grow with follicular wall getting thinner, the follicular fluid increasing, and follicular cavity expanding. The follicles extend to the cortex of the whole ovary and protrude from the surface of the ovary. The sizes of the mature follicles from different animals vary greatly. The diameter of mature follicles of the cattle is 10-14 mm, the pig is 8-12 mm, the sheep is 5-10 mm, and the goat is 7-10 mm. In an estrus cycle, multiple primordial follicles of multiparous animals can develop into mature follicles at the same time, while only one follicle of the uniparous animal can mature and ovulate generally.

4.4 Follicular atresia and degeneration

Before birth, many primordial follicles exist in the ovary, but only a few of them can mature and ovulate, and most of them degenerated. The younger the animal is, the more severe the occurrence of follicular atresia is. The cause of follicular atresia may be attributed to insufficient secretion of FSH in the anterior pituitary or

叶 FSH 的分泌不足或卵泡对 FSH 的反应性降低有关。 | decreased reactivity of the follicles to FSH.

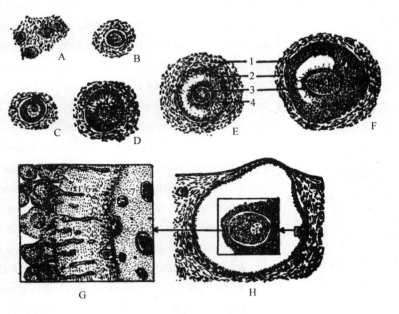

图 3-6 哺乳动物卵泡发育模式图

Figure 3-6 Pattern of follicular development in mammals

A. 原始卵泡 B. 初级卵泡 C. 次级卵泡 D. 三级卵泡 E. 出现新月形腔隙的三级卵泡
F. 出现卵丘的三级卵泡 G. 卵黄膜的微绒毛部分伸向透明带 H. 成熟卵泡

A. primordial follicle B. primary follicle C. secondary follicle D. tertiary follicle
E. tertiary follicle with crescent follicular cavity F. tertiary follicle with the cumulus
G. microvilli of yolk membrane extending to zona pellucida H. mature follicle

1. 卵泡外膜 2. 颗粒层 3. 透明带 4. 卵丘 5. 颗粒层细胞 6. 透明带 7. 卵黄
1. follicular adventitia 2. granular layer 3. zona pellucida 4. cumulus 5. granulosa cell
6. zona pellucida 7. ooplasm

卵泡的生长发育
Growth and development of follicular

（五）排卵

成熟卵泡破裂、释放卵子的过程，称为排卵。生长发育到一定阶段，卵泡成熟破裂，随着卵泡液的流出，卵子与卵丘脱离而被排出卵巢外，由输卵管伞部接纳。

1. 排卵的类型

（1）自发性排卵。卵巢上的成熟卵泡自行破裂排卵并自动形成黄体。这种类型又有两种情况：一是发情周期中黄

4.5 Ovulation

The mature follicles rupture, thus freeing the ovum in a process called ovulation. At a certain stage of growth and development, the follicle matures and ruptures, then frees as the follicular fluid drains. The ovum separated from the cumulus is discharged out from the ovary, and enters the fimbriae of the oviduct.

4.5.1 Types of the ovulation

（1）Spontaneous ovulation. The mature follicle on the ovary ruptures and spontaneously forms the corpus luteum. Firstly, the function of the corpus luteum can

体的功能可以维持一定时期，且具有功能性，如牛、猪、马、羊等属于此种类型；二是除非交配（刺激），否则形成的黄体是非功能性的，即不分泌孕酮，鼠属于这种类型。

（2）诱发性排卵。又称为刺激性排卵，即卵泡发育成熟后，必须通过交配或其他途经使子宫颈受到机械性刺激后才能排卵，并形成功能性黄体。兔、骆驼、猫等属于诱发性排卵动物，它们在发情季节中，卵泡有规律地陆续成熟和退化，如果交配（刺激），随时都有成熟的卵泡排卵。

2. 排卵时间

排卵是成熟卵泡在LH峰作用下产生的，各种动物的排卵时间因动物种类、品种、个体、年龄、营养状况及环境条件等不同而异。常见家畜的排卵时间如下：牛在发情结束后8～12h，猪在发情开始后16～48h，羊在发情结束时，兔在交配刺激后6～12h。

3. 排卵数目

牛、马等大家畜一般为1枚，个别的可排2枚；绵羊1～3枚；山羊1～5枚；猪10～25枚；兔5～15枚。

（六）黄体形成与退化

成熟卵泡排卵后形成黄体，黄体分泌孕酮作用于生殖道，使之向妊娠方向变化。如未受精，一段时间后黄体退化，开始下一次的卵泡发育与排卵。

成熟卵泡破裂排卵后，卵泡腔产生

be maintained in a certain period of estrus cycle, such as cows, sows, mares, ewes, etc. Secondly, unless mating (stimulation), the formed corpus luteum would lose its function, which means the progesterone is not secreted by the corpus luteum, such as the case in rats.

(2) Induced ovulation. It is also known as stimulating ovulation. After follicles maturation, the cervix must be mechanically stimulated by mating or other ways to ovulate and form functional corpus luteum. Induced ovulation occurs in animals such as rabbits, camels, cats and so on. During the estrus season, their follicles mature and degenerate regularly and gradually. However, if mating (stimulation) occurs, mature follicles will ovulate at any time.

4.5.2　The time of ovulation

Ovulation occurs under the LH peak stimulated by mature follicles. The time of ovulation varies according to animal species, breed, individual, age, nutritional status and environmental conditions. The time of ovulation is 8-12 hours after end of estrus in cows, 16-48 hours after onset of estrus in sows, near the end of estrus in ewes, and 6-12 hours after the stimulation of mating in rabbits.

4.5.3　The number of ovulation

The number of ovulation is one generally, occasionally two in large domestic animals such as cows and mares, 1-3 in ewes, 1-5 in goats, 10-25 in sows, and 5-15 in rabbits.

4.6　The formation and degeneration of the corpus luteum

After ovulation of mature follicles, it forms the corpus luteum, which secretes progesterone and stimulates the reproductive tract to prepare for pregnancy. In non-fertilized animals, the corpus luteum gradually degenerates and starts the next follicular development and ovulation.

After the rupture and ovulation of the mature fol-

负压，卵泡膜血管破裂流血，并充溢于卵泡腔内形成血凝块，称为红体。此后颗粒细胞在LH作用下增生变大，并吸取类脂质而变成黄体细胞。同时卵泡内膜血管增生分布于黄色细胞团中，卵泡膜的部分细胞也进入黄色细胞团，共同构成了黄体。黄体在排卵后7～10d（牛、羊、猪）或14d（马）发育至最大体积。

黄体是一种暂时性的分泌组织，主要作用是分泌孕酮。在发情周期中，如果雌性动物没有妊娠，所形成的黄体在黄体期末退化，这种黄体称为周期性黄体。周期性黄体通常在排卵后维持一定时间才退化，退化时间牛为14～15d，羊为12～14d，猪为13d，马为17d。如果雌性动物妊娠，则黄体存在的时间长，体积也增大，这种黄体称为妊娠黄体。妊娠黄体分泌孕酮以维持妊娠需要，直至妊娠结束时才退化。但马、驴的妊娠黄体退化较早，一般在妊娠后160d即开始退化，以后靠胎盘分泌孕酮维持妊娠。

五、发情生理

雌性动物生长发育到一定年龄后，受下丘脑-垂体-卵巢轴调控，每隔一定时间，卵巢上就有卵泡发育，逐渐成熟而排卵，此时雌性动物的精神状态和生殖器官及行为都发生一系列变化，如兴

licle, the follicular cavity produces the negative pressure, with the blood vessels on the follicular membrane rupture and blood, and the cavity of the ruptured follicle fills with a blood clot, which is called corpus hemorrhagic. Thereafter, the granulosa cells proliferate under the action of LH, and absorb lipoid to develop into the luteal cells. At the same time, the vascular proliferation on follicular endometriosis distribute into the yellow cell mass, and some cells of the follicular theca also enter the yellow cell mass, which join together to form the corpus luteum. The corpus luteum develops to the maximum volume in 7-10 days (cow, ewe and sow) or 14 days (mare) after ovulation.

The corpus luteum is a temporary secretory tissue. Its main function is to secrete progesterone. During an estrous cycle, if the female animal does not have a pregnancy, the formed corpus luteum degenerates at the end of the luteal phase, which is called periodic corpus luteum. The periodic corpus luteum usually lasts for a time after ovulation, and then degenerate. The time of the regression is 14-15 days in cows, 12-14 days in ewes, 13 days in sows, 17 days in mares. If the female animal is pregnant, the corpus luteum lasts for a long time and enlarges in volume, this cropus luteum is called pregnant corpus luteum. The pregnant corpus luteum secretes progesterone to maintain pregnancy, and it does not degenerate until the end of pregnancy. However, the corpus luteum degenerates earlier in horse and donkey. It usually begins to degenerate in 160 days after pregnancy, afterwards the pregnancy is maintained by the progesterone secreted by the placenta.

5 Estrus physiology

When the female animal grows and develops to a certain age, at the regular intervals, the follicles on the ovary matures gradually to ovulate, and this process is regulated by the hypothalamic-pituitary-ovarian axis. At this time, the mental state, reproductive organs and behaviors of the female animals all undergo a series of

奋不安，食欲减退、外阴肿胀、阴道黏液增多，有强烈的求偶行为。

（一）发情周期

雌性动物在初情期以后，表现出周而复始的性活动周期。即在生理或非妊娠条件下，母畜每间隔一定时期均会出现一次发情，通常把一次发情开始至下次发情开始，或一次发情结束至下次发情结束所间隔的时期，称为发情周期。牛、猪、山羊的发情周期平均为 21d，绵羊为 16~17d。根据机体所发生的一系列生理变化，发情周期多采用四期分法和二期分法（图 3-7）。

changes, such as excitement and unrest, appetite loss, vulvar swelling and mucus increasing and display strongly courtship behaviors.

5.1 Estrus cycle

Females show a cycle of sexual activity after puberty. Under physiological or non-pregnant conditions, an estrus occurs periodically at regular time intervals in females. Usually, the interval period, from the onset of an estrus to the onset of next estrus, or from the end of an estrus to the end of next estrus, is called an estrous cycle. The estrous cycle is 21 days on average in cows, sows and female goats, and 16-17 days in ewes. Based on a series of physiological changes, an estrous cycle is usually divided into various phases using four-phase or two-phase methods generally (Figure 3-7).

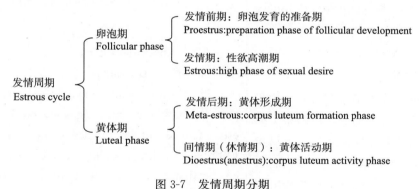

图 3-7　发情周期分期

Figure 3-7　Phases of the estrous cycle

（二）发情持续期

发情持续期是指母畜从发情开始到发情结束所持续的时间，相当于发情周期中的发情期。各种母畜的发情持续期为：牛 1~2d，羊 1~1.5d，猪 2~3d，马 4~7d。由于季节、饲养管理水平、年龄及个体条件的不同，母畜发情持续期的长短也有所差异。

5.2 Duration of estrus

The duration of estrus is the length of time from the onset of estrus to the end of estrus, equivalent to estrus phase in the estrus cycle. The duration of estrus in various females is as follows: 1-2 days in cows, 1-1.5 days in ewes, 2-3 days in sows, and 4-7 days in mares. The variable length of the duration of the estrous in various females depends on differences in season, feeding management, age and individual conditions.

（三）发情季节

雌性动物发情主要受神经内分泌调控，但也受外界环境条件的影响，季节变化是影响发情周期的重要环境因素。有些动物如马、绵羊、犬等，一年中只在特定季节才表现出发情。动物的发情可分为季节性发情和全年发情两种类型。季节性发情又分为季节性多次发情和季节性单次发情两种。

季节性多次发情是指在发情季节有多个发情周期。如马、绵羊在春季或秋季发情时，如果没有配种或配种后未受胎，可出现多次发情。犬的发情季节为春、秋两季，但在每个发情季节内只有一个发情周期。全年多次发情是指雌性动物在一年四季都可以出现发情并可配种，牛、猪等属此类型。

（四）异常发情

1. 安静发情

安静发情又称为隐性发情，是指母畜发情时外部表现不明显，但卵巢上有卵泡发育、成熟并排卵。常见于产后第一次发情以及营养不良的牛、马和羊。雌激素分泌不足或体内缺乏孕酮均可引起安静发情。

2. 短促发情

短促发情指母畜发情持续时间短，往往错过配种时机。其原因可能是神经内分泌系统的功能失调，发育的卵泡很快成熟破裂排卵，缩短了发情期，也可能是由于卵泡突然停止发育或发育受阻而引起。

3. 断续发情

断续发情即母畜发情延续很长，且

5.3 Estrus season

The estrus in females is mainly regulated by neuroendocrine, and also affected by external environment. The seasonal changes are important factors affecting the estrus cycle. The estrus in some species, such as horses, sheep and dogs, occurs in a certain season. There are seasonal estrus and annual estrus. Seasonal estrus includes seasonal poly-estrus and seasonal mono-estrus.

The seasonal poly-estrus represents the poly-estrus cycles during the estrus season. For example, while a horse or sheep is estrus in spring or autumn, if they are not mated or un-pregnant, poly-estrus will occur. The estrus seasons in dogs are spring and autumn, but only one estrus cycle occurs in each estrus season. Poly-estrus yearly means that the females can be estrus and mated all year round, such as cows and sows.

5.4 Abnormal estrus

5.4.1 Silent estrus

Silent estrus, also known as recessive estrus, represents that the external signs are not obvious in the estrus female, but the follicles on the ovary mature, develop and ovulate normally. It is commonly observed on the first postpartum estrus or malnourished cows, mares and ewes. The silent estrus is caused by insufficient secretion of estrogen or loss of progesterone in vivo.

5.4.2 Short estrus

The short estrus represents a short duration of estrus in females. It is easy to miss the mating time. The reason may be the dysfunction of the neuroendocrine system, resulting that the developing follicles quickly mature, rupture and ovulate, thus shortening the estrus period. It also may be caused by sudden cessation of follicular development or obstruction of follicular development.

5.4.3 Intermittent estrus

This form of estrus in the females lasts for a long

发情时断时续。多见于早春或营养不良的母马。其原因是卵泡交替发育的结果。

4. 持续发情

持续发情是慕雄狂的症状之一,表现为持续强烈的发情行为。患慕雄狂的母牛,表现为极度不安,大声哞叫,频频排尿,追逐爬跨其他母牛,产奶量下降,食欲减退,身体消瘦,外形往往具有雄性特征,如颈部肌肉发达等。

5. 孕后发情

孕后发情又称假发情,是指妊娠母畜仍有发情表现。母牛在妊娠最初3个月内,常有3%～5%的母牛发情,绵羊孕后发情可达30%。孕后发情发生的主要原因是激素分泌失调,即妊娠黄体分泌孕酮不足,而胎盘分泌雌激素过多所致,容易引起流产,称之为"激素性流产"。生产上常常会因为误配而造成流产,因此要认真区分。

(五)产后发情

产后发情是指母畜分娩后的第一次发情。奶牛一般在产后25～30d发情,多数表现为安静发情,产后40～50d正常发情。母猪一般在分娩后3～6d内可出现发情,但不排卵。在仔猪断奶后一周之内,80%左右的母猪出现第一次正常发情。母羊大多在产后2～3个月发情。母马往往在产驹后6～12d发情,一般发情表现不太明显,甚至无发情表现,但有卵泡发育且可排卵,若配种可受孕。

time, characterized by an intermittent process. It is common in early spring time and in malnourished mares, which can be attributed to the alternate development of follicles.

5.4.4 Continuous estrus

The continuous estrus is one of the symptoms of nymphomania, characterized by strongly persistent estrus behavior. The cows suffering from nymphomania are characterized by extremely uneasy, screaming loudly, frequent micturition, chasing and mounting other cows, reduction of milk production, loss of appetite, weight loss, and appearance of maleness involving muscular neck etc.

5.4.5 Estrus after pregnancy

It is also known as false estrus, which means that the pregnant females still have estrus behavior. The estrus often occurs in 3%-5% of cows in the first three months after pregnancy, and in 30% of ewes after pregnancy. The main cause of this phenomenon may be the imbalance of hormone secretion. Such as, insufficient progesterone secreted by pregnant corpus luteum and excessive estrogen secreted by the placenta, can cause abortion easily, which is called "hormonal abortion". In practical production, it is necessary to distinguish the real estrus from the false ones after pregnancy to avoid abortion.

5.5 Postpartum estrus

It is the first estrus in females after delivery. The estrus generally occurs in cows in 25-30 days after delivery, mostly in the form of silent estrus, and it is seen as normal estrus after 40-50 days after delivery. Sows usually display estrus within 3-6 days after delivery without ovulation. About 80% of them reach the first estrus normally within one week after weaning of piglets. Estrus of ewes usually occurs within 2-3 months after delivery. Mares often reach the estrus within 6-12 days after delivery, with inconspicuous or no estrus behavior, however the follicular development and ovulation still occur, and they can pregnant after mating.

（六）乏情

乏情是指达到初情期的雌性动物长期不发情，卵巢无周期性的功能活动，而是处于相对静止状态。引起乏情的因素很多，有季节性、生理性和病理性等。

1. 季节性乏情

季节性乏情的动物在非繁殖季节，卵巢和生殖道处于静止状态。季节性乏情的时间因畜种、品种和环境而异。马多为短日照的冬春季乏情，绵羊的乏情往往发生于长日照的夏季。

2. 生理性乏情

生理性乏情包括泌乳性乏情、妊娠期乏情和衰老性乏情等。猪是最常见的泌乳性乏情动物，泌乳期间发情和排卵受到抑制，一般在仔猪断乳后才出现发情。妊娠期间由于卵巢上存在妊娠黄体，可以分泌孕酮而抑制发情。妊娠期乏情是保证胚胎正常发育的生理现象。动物因衰老使下丘脑-垂体-性腺轴的功能减退，导致垂体促性腺激素分泌减少，或卵巢对这些激素的反应性降低。

3. 病理性乏情

营养不良、应激反应、卵巢机能疾病（如持久黄体、黄体囊肿等）均会抑制发情。

5.6 Anestrus

Being anestrus means that the females reaching puberty can not begin the process of the estrus for a long time, and the ovary is in a relatively static state. There are many factors causing anestrus, such as seasonality, physiology and pathology reasons etc.

5.6.1 Seasonal anestrus

The ovary and reproductive tract of the seasonal anestrus animals are at rest in the non-breeding season. The time of seasonal anestrus varies depending on species, breed and environment. For example, the anestrus in mares occur in winter and spring with short-daylight, while the anestrus in ewes often occur in summer with long-daylight.

5.6.2 Physiological anestrus

Physiological anestrus includes lactational anestrus, pregnant anestrus and aging anestrus. Sows are the most common lactational anestrus animals. During lactation, estrus and ovulation are inhibited. Estrus usually occurs after weaning. The females can secret progesterone to inhibit estrus in pregnancy due to the presence of gestational corpus luteum on the ovary. Anestrus in pregnancy is a physiological phenomenon that ensures the normal development of the embryo. Due to aging, the function of the hypothalamic-pituitary-gonadal axis is reduced in animals, resulting in a decrease in the secretion of pituitary gonadotropins, or a decrease in the ovarian response to these hormones.

5.6.3 Pathological anestrus

Malnutrition, stress and ovarian dysfunction (such as persistent corpus luteum and corpus luteum cysts etc.) can inhibit estrus.

项目三 发情鉴定与输精
Project III Estrus Identification and Insemination

任务1 母牛的发情鉴定与输精
Task 1 Estrus Identification and Insemination of Cows

任务描述
Task Description

人工授精是世界养牛业的通用繁殖方式,也是牛场生产管理的关键环节。作为一名牛场配种员,不仅要具备娴熟的专业技术和丰富的生产经验,还要有很强的责任心。如何鉴定母牛是否发情?如何规范完成输精?

Artificial insemination is widely used in dairy farms all over the world, and it is a key link in management. As a breeder, we are supposed to not only have the professional technology and abundant experience, but also have a strong sense of responsibility. How to identify whether cows are estrus or not? How to standardize the insemination process?

任务实施
Task Implementation

一、母牛的发情鉴定
1 Estrus identification of cows

(一)外部观察法
1.1 External observation

外部观察法主要是通过观察母牛的外部表现和精神状态来判断其发情情况。例如发情母牛表现为兴奋不安,食欲减退,有强烈的求偶行为,外阴充血、肿胀、湿润有透明黏液,产乳量下降等。

External observation is mainly used to judge the estrus status of cows by observing their external performance and mental state. For example, estrus cows show excitement, loss of appetite, strong courtship behaviors, pudendal congestion, swelling and moist with transparent mucus, milk production decline, etc.

1. 准备工作

将母牛放入运动场或在牛舍内观察,一般早晚各一次,可利用照相机或摄像机。

1.1.1 Preparations

Observe the cows in the stadium or cowshed, usually once in the morning and once in the evening with cameras.

2. 检查方法

主要通过观察母牛的爬跨情况、外阴部的肿胀程度及黏液的状态,进行综合分析判断。

1.1.2 Examination method

Make a comprehensive analysis and judgment by observing the mounting behaviours, along with vulva swelling and mucus status.

3. 结果判定

发情母牛表现为食欲下降,兴奋不安,大声哞叫,四处走动,爬跨或接受爬跨(图3-8)。外阴肿胀,阴道黏膜潮

1.1.3 Results judgement

Estrus cows show decreased appetite, increased excitement, loud barking, walking around, mounting behavious (Figure 3-8). The vulva is swollen, the vaginal mucosa is flushed, the uterus cervix is opened,

红，子宫颈开张，外阴流出牵缕样或玻璃棒状黏液（图3-9）。

mucus outflow is like a stretch or glass rod (Figure 3-9).

图 3-8　母牛发情时的爬跨行为

Figure 3-8　Mounting behavior of estrous cows

图 3-9　发情母牛阴户排出的黏液

Figure 3-9　Mucus excreted from the vulva of estrous cows

外部观察法

External observation

（二）直肠检查法

1. 准备工作

将待检母牛牵入保定栏内，牛尾拉向一侧，使肛门充分外露。检查人员指甲剪短磨光，带上长臂手套，并涂抹少量的润滑剂。

2. 检查方法

检查人员站在母牛的正后方，手指并拢呈锥形，旋转且缓慢伸入肛门，进入直肠并将宿便清出。然后，将手掌向骨盆腔底部下压找到子宫颈，手轻握子

1.2　Rectal examination

1.2.1　Preparations

The assistant leads the cow into the fixer and pulls the tail to one side, so as to let the anus fully exposed. The inspectors cut and polish nails, wear gloves with long arms and apply a small amount of lubricant.

1.2.2　Examination method

The inspector should stand right behind the cow, keep fingers together in a conical shape, rotate and extend slowly into the anus. Clean up the stool in the rectum, and then press the palm down to the bottom of the pelvic cavity to find the cervix. After finding the

宫颈向前滑动到角间沟，继续向前找到卵巢，触摸其大小、形状、质地以及卵泡的发育情况（图3-10）。

cervix, gently hold the cervix and slide forward into the angular sulcus, continue to find out ovaries and feel the size, shape, texture, and follicle development (Figure 3-10).

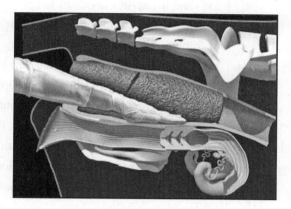

图 3-10　直肠检查示意
Figure 3-10　Rectal examination

直肠检查法
Rectal examination

3. 结果判定

母牛在间情期，一侧卵巢较大，能触到一个枕状的黄体突出于卵巢的一端；当母牛进入发情期以后，则能触到有一个黄豆大的卵泡存在，这个卵泡由小到大，由硬到软，由无波动到有波动。由于卵泡发育，卵巢体积变大，直肠检查时容易摸到。为了便于区分，确定最佳的输精时间，通常将卵泡发育分为四个时期（图3-11）。

1.2.3　Results judgment

During diestrus, cows have larger ovaries on one side than the other, and a pillow-shaped corpus luteum could be touched protruding from one end of the ovary. During estrus, a large follicle likes a soybean could be felt, which will change from small to large, hard to soft, and without wave to with wave at the end. Ovarian volume becomes larger due to follicular development, which is easily palpable during rectal examination. In order to distinguish and determine the best time for insemination, follicular development is usually divided into four phases (Figure 3-11).

A　　　　　　　　　B　　　　　　　　　C　　　　　　　　　D

图 3-11　牛卵泡发育模式
Figure 3-11　Model of follicular development in cow
A. 卵泡出现期　B. 卵泡发育期　C. 卵泡成熟期　D. 排卵期
A. follicular appearance　B. follicular development　C. follicular maturation　D. ovulation

第一期（卵泡出现期）：卵泡稍增大，直径为 0.5~0.75cm，直肠触诊为一硬性隆起，波动不明显。此期母牛已有发情表现，约持续 10h。

第二期（卵泡发育期）：卵泡直径增大到 1~1.5cm，并突出于卵巢表面，呈小球状，波动明显。此期母牛处于外部表现的盛期，持续 10~12h。

第三期（卵泡成熟期）：卵泡不再继续增大，但卵泡液增多，卵泡壁变薄，紧张度增强，直肠触诊时有"一触即破"的感觉，似熟葡萄。

第四期（排卵期）：卵泡成熟破裂，卵泡液流出，卵巢上留下一个小的凹陷。排卵后 6~8h 可摸到肉样感觉的黄体，其直径为 0.5~0.8cm。

母牛最佳的输精时间一般选择在卵泡成熟期之后，此期母牛性欲减退，没有明显的发情征状。

（三）涂蜡笔法

涂蜡笔法鉴定发情母牛，涂抹部位为尾椎上面，从尾部到十字部，长度 30~40cm（图 3-12）。每天涂蜡笔 1~2 次，以早晨为佳。

发情母牛接受其他母牛爬跨后，尾部毛发被压，上面的蜡笔涂料被摩擦掉，或者被其他母牛腹部黏附的牛粪污染，颜色变浅、变深。未发情母牛不接受其他母牛爬跨，尾部毛发直立或高耸，蜡笔涂料新鲜，与新涂抹的保持一致。

（四）计步器识别法

计步器识别法是基于发情动物的运动量要显著高于未发情动物的原理，

Phase Ⅰ (follicular appearance): The follicle is enlarged slightly, with a diameter of 0.5-0.75 cm. The rectal palpation indicates a rigid bulge with no obvious fluctuation. The cows in this period show signs of estrus, and it lasts for about 10 hours.

Phase Ⅱ (follicular development): The diameter of follicles is 1.0-1.5 cm, protruding on the ovarian surface and showing a globular shape with obvious fluctuation. The cows in this period are at the peak of external performance, which lasts about 10-12 hours.

Phase Ⅲ (follicular maturation): The size of follicles stops growing, but follicular fluid increases. The follicular wall becomes thinner, tension increases and rectal palpation indicates a "break-upon-touching" feeling. Mature follicle shape likes a ripe grape.

Phase Ⅳ (ovulation): Matured follicular ruptures, and follicular fluid flows out, leaving a small depression on the ovary. Meat-like corpus luteum can be felt 6-8 hours after ovulation, and its diameter is about 0.5-0.8 cm.

The best insemination time for cows is usually after Phase Ⅲ, during which the cow's sexual desire decreases and has no obvious estrus symptoms.

1.3 Painting method

To identify estrus cows, the area above the cow's caudal vertebrae, from the tail to the cross, can be painted and the length should be about 30-40 cm (Figure 3-12). Repeat the painting 1-2 times a day, preferably in the morning.

If the cows are estrus, they would accept mounting and the painting area would be rubbed off or contaminated by the dung adhering to the abdomen of other cows. If the cows are diestrus, they would not accept mounting, their tail hair will be upright or towering, so the painting area stays fresh and be consistent with the new painting.

1.4 Pedometer recognition

The method of pedometer recognition is based on the principle that the movement of the estrus animal is

通过对母牛运动量的监测辅助鉴定发情母牛。计步器安装简单，灵敏度高，传感器识读率高，发情监测准确，系统会根据每天的活动量自动判断母牛的发情状态，以此大大简化牛场繁殖人员的工作。

significantly higher than that of the non-estrus animal. The pedometer has the advantages of simple installation, high sensitivity, high sensor and high accuracy of estrus detection. The system can automatically judge the estrus state of cows according to daily activity, which greatly simplifies the work of cow breeders.

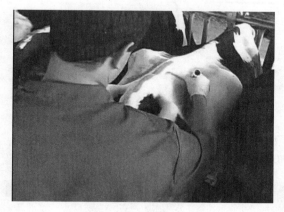

图 3-12　涂蜡笔部位
Figure 3-12　Painting area

二、母牛的输精

（一）准备工作

1. 母牛

母牛经发情鉴定后，将其保定，清洗、消毒外阴，尾巴拉向一侧，使外阴部充分展露。

2. 器械准备

输精器械使用前必须彻底洗净、消毒。牛用各种输精枪及外套管见图 3-13。外套管独立包装，前端圆滑，输精时可避免划伤子宫颈和精液倒流。

3. 精液准备

将解冻的细管精液棉塞端插入输精枪推杆 0.5cm，剪掉细管封口部，装上钢管套（图 3-14）。

2　Insemination of cows

2.1　Preparations

2.2.1　Cow

Pull the estrus cows into the fixer, clean and sterile its vulva, pull the tail to one side so as to fully exposed the vulva.

2.2.2　Instruments preparation

The instruments used for insemination must be thoroughly cleaned and disinfected before using. Various semen injectors and external cannulas for cows are shown in Figure 3-13. The outer cannula is packed independently and the front end of which should be smooth, thus to avoid cervical scratches and semen reflux during insemination.

2.1.3　Semen preparation

Thaw the frozen semen tubule and then insert the cotton end into the push rod of semen injector at about 0.5 cm. Cut off the sealing part of the tubule and install the steel tube cover(Figure 3-14).

4. 输精人员准备

输精人员穿好工作服，将指甲剪短磨平，手及手臂清洗后消毒。伸入直肠的手臂要戴一次性直肠检查手套并涂抹润滑剂。

2.1.4 Inseminator preparation

Inseminators should wear overalls, cut and polish their nails. Hands and arms should be cleaned and disinfected. The arm extending into the rectum should wear disposable gloves and apply lubricant.

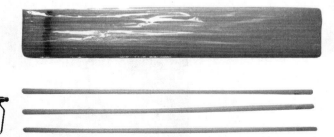

图 3-13　牛用输精枪及外套管
Figure 3-13　Semen injectors and external cannulas for cow

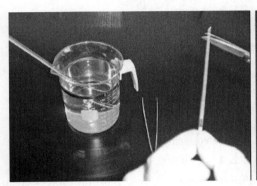

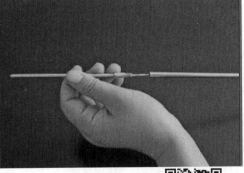

图 3-14　精液准备
Figure 3-14　Semen preparation

精液准备
Preparation of semen

（二）输精操作

目前，牛的输精普遍采用直肠把握子宫颈输精法（图 3-15）。

输精人员站在母牛正后方，一只手戴长臂手套，五指并拢呈锥形伸入直肠内把握住子宫颈，另一只手持输精器，先向斜上方伸入阴道内 5~10cm，避开尿道开口，然后两手协同配合，把输精器插入子宫颈内 3~4 个皱褶处，缓慢注入精液。输精完毕后，先抽出输精器，然后撤出手臂。

2.2 Insemination

At present, the method of rectal control of cervical insemination is widely used in cows (Figure 3-15).

The inseminator stands right behind the cow, wears long arm glove on one hand, puts five fingers together and cones into the rectum to grasp the cervix, and the other hand holds the inseminating syringe. First, insert the inseminating syringe into the vagina, which is about 5-10 cm above the oblique direction, avoiding to enter the urinary meatus. Then with two hands cooperation, insert the inseminating syringe into

项目三　发情鉴定与输精

Project Ⅲ　Estrus Identification and Insemination

输精过程中，输精器不要握得太紧，要随着母牛的摆动而灵活伸入；保持子宫颈呈水平状态（图 3-16）；避免盲目用力插入，防止损伤生殖道黏膜；输精的原则为"轻插、适深、缓注、慢出"，防止精液倒流。

直肠把握子宫颈输精法具有安全、母牛无痛感、受胎率高等优点，并可隔着直肠触摸母牛子宫和卵巢的变化判断发情和妊娠情况，防止误配或流产。但此法初学者不容易掌握，如果操作过程中把握子宫颈的手掌位置不准，会导致输精枪插入过浅而降低受胎率。

the cervix at 3-4 folds. While in the right position, slowly inseminate the semen. After insemination, pull out the syringe first and then the arm.

During the insemination, the inseminating syringe should not be held too tightly, it should be extended flexibly with the moves of the cow, while keeping the level position of the cervix (Figure 3-16). Don't insert the syringe blindly and forcefully to avoid genital tract mucosal damage. The principle of insemination is light insertion, moderate depth, slow injection, slow pull out and prevent reflux of semen.

Rectal control of cervical insemination has the advantages of safety operation, painless of cows, high pregnancy rate. What's more, it can be used to judge estrus and pregnancy stage by touching the uterus and ovaries of cows, so as to prevent mismating or abortion. However, this method is not easy for beginners to master. If the position of the cervix is not accurate during the operation, the insertion of the syringe would not be in the correct position which would reduce the fertility rate.

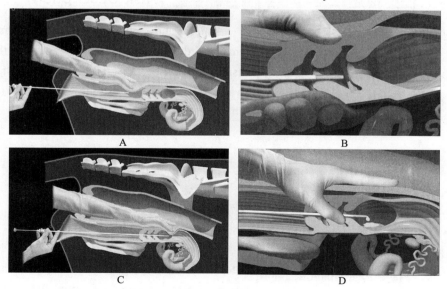

图 3-15　直肠把握子宫颈输精

Figure 3-15　Rectal control of cervical insemination

A. 输精器伸入阴道　B. 输精器进入子宫皱褶　C. 输精器进入子宫颈口　D. 缓慢注入精液

A. insert syringe into vagina　B. insert syringe into uterine fold　C. insert syringe into cervical orifice

D. inject semen slowly

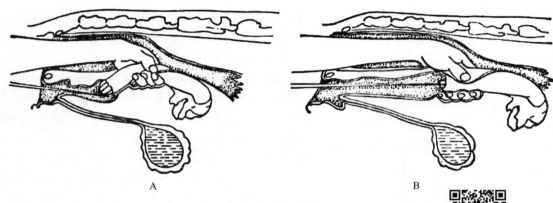

A   B

图 3-16 把握子宫颈的操作

Figure 3-16 The operation of cervix control

A. 错误操作　B. 正确操作

A. incorrect operation　B. correct operation

直肠把握子宫颈输精

Rectal control of cervical insemination

任务 2　母羊的发情鉴定与输精
Task 2　Estrus Identification and Insemination of Ewes

任务描述 / Task Description

适时而准确地把优质精液输送到发情母羊的子宫颈口内，是保证母羊受胎、妊娠和产羔的关键。作为一名羊场配种员，如何进行母羊的输精呢？

Timely and accurate delivery of high-quality semen to the cervix of estrous ewes is the key to ensure the conception, pregnancy and lambing of ewes. As a sheep farm breeder, how can you successfully complete the insemination of ewes?

任务实施 / Task Implementation

一、母羊的发情鉴定
1　Estrus identification of ewes

1. 外部观察法
1.1　External observation

发情母羊精神兴奋不安，不时地高声"咩"叫，并接受其他羊的爬跨。同时，食欲减退、频频排尿，用手按压母羊背部，母羊站立不动、摆尾，有交配欲。发情母羊的外阴部及阴道充血、肿胀、松弛，并有黏液流出。

The estrus ewes are agitated, crying loudly and accepting the humping of other sheep. What's more, loss of appetite, frequent urination, standing still while pressing the back with the tail wagging, with mating desire are also commonly observed. The vulva and vagina of estrus ewes are congested, swollen and relaxed, and with mucus flowing out.

2. 试情法

试情公羊的头数为母羊数的3%~5%。在试情公羊的腹部采用标记装置或在胸部装上染料囊（图3-17），每日1次或早晚各1次定时放入母羊群中，如果母羊发情并接受公羊爬跨，便将颜色印在母羊背部上。试情结束后，最好选用另一头试情公羊，对挑出的全部发情母羊重复试情一次，准确率达90%以上。

1.2 Testing method

The rate of rams is 3%-5% of ewes. Under the abdomen of the test ram, a labeling device or a dye bag is used (Figure 3-17). Send the rams into the ewes group once a day or once in the morning and once in the evening. If the ewe is during estrus, it would accept the mounting of the ram, and the color will leave a trace on the back of the ewe. At the end of the test, it is better to repeat the test once for all selected ewes using another testing ram, and the accuracy rate is greater than 90%.

图 3-17　羊的试情兜布
Figure 3-17　Testing cloth

3. 阴道检查法

阴道检查法是通过开膣器检查母羊阴道内变化来判定母羊是否发情。该法操作简单、准确率高，但工作效率低，适于小规模饲养户应用。

检查时，将待检母羊保定，清洗外阴，擦干，用酒精棉球消毒。将开膣器前端闭合，缓慢插入母羊阴道，轻轻打开前端，借助光线观察阴道内的变化。发情母羊阴道黏膜潮红、充血、表面光亮湿润，有透明黏液渗出，子宫颈口松弛、开张、呈深红色。未发情母羊阴道

1.3 Vaginal examination

Vaginal examination is to determine whether the ewe is estrus by examining the changes in the vagina of the ewe with an opener. The method is simple and high in accuracy, but it is inefficient and only suitable for small-scale feeders.

During examination, fix the ewe, clean the vulva, dry it and disinfect with alcohol. Close the fore-end of the opener and insert it into the vagina of the ewe slowly. After entering in the vagina, open the opener gently and observe the vagina by light. The vaginal mucosa of estrus ewe is flushed, congested, bright and moist, with transparent mucus exudation, and the cervical orifice is loose, open and dark red. The vaginal

黏膜苍白、干涩，开膛器伸入有阻力，子宫颈口关闭。检查后稍微合拢开膛器前端，抽出。

mucosa of the diestrus ewe is pale and dry, and the opener is obstructed and the cervical orifice is closed. After the examination, slightly close the fore-end of the opener and pull it out.

二、母羊的输精

母羊输精有开膛器输精、阴道输精和腹腔镜子宫内输精等方法，目前生产上普遍采用开膛器输精法。

助手倒提母羊，输精员将消毒开膛器涂抹润滑剂，旋转插入母羊阴道内，打开开膛器，找到子宫颈外口，另一只手持输精器沿开膛器插入子宫颈口内0.5～1cm缓慢注入精液（图3-18），然后撤出输精器，并将半开半闭状态的开膛器慢慢取出。最后输精员轻拍母羊的腰背部，防止精液倒流。

2 Insemination of ewes

There are several methods for ewe insemination, such as opener insemination, vaginal insemination and laparoscopic insemination. At present, opener insemination is widely used in ewe production.

When the assistant lifts the ewe upside down, the inseminator rubs the sterilized opener with lubricant, rotates it into the vagina, opens the opener and finds the cervix. Insert the insemination syringe along the opener into the cervix at 0.5-1 cm and slowly inject the semen (Figure 3-18). Withdraw the syringe first and then the opener. Finally, the inseminator pats the back of the ewe lightly to prevent the reflux of semen.

图 3-18 羊的开膛器输精
Figure 3-18 Insemination of ewe with opener

开膛器输精
Opener insemination

任务 3 母猪的发情鉴定与输精
Task 3 Estrus Identification and Insemination of Sows

任务描述

发情鉴定是母猪饲养管理工作中的重中之重，熟悉母猪的发情规律，掌握正确的发情鉴定方法，以便确定配种适

Task Description

Estrus identification is the most important task in the breeding and management of sows. It is necessary to understand the estrus patterns and master the estrus

期，从而提高受胎率和产仔数。作为一名猪场配种员，如何准确鉴定母猪是否发情？如何确定最佳配种时间？母猪输精时需注意哪些事项？如何防止精液倒流？

identification of sows, so as to determine the suitable time for insemination and improve the conception rate and litter size. As a pig breeder, how to accurately identify whether the sows are estrus? How to determine the best time for insemination? What should we pay attention to during insemination? And How to prevent semen reflux?

任务实施

一、母猪的发情鉴定

1. 外部观察法

母猪开始发情时对周围环境十分敏感，兴奋不安，食欲减退，两耳耸立，东张西望，外阴明显充血、肿胀，阴唇黏膜随着发情盛期的到来，变为淡红色或血红色，黏液量多而稀薄，性欲趋向旺盛。随后，母猪食欲回升，表现呆滞，阴门变为淡红、微皱、稍干，黏液由稀转稠，愿意接受爬跨（图3-19），此时母猪进入发情末期，是配种的最佳时期。

Task Implementation

1 Estrus identification of sows

1.1 External observation

Sows are very sensitive to the environment when they start estrus. They are very excited and restless, showing appetite decreasing, two ears standing up and looking around. The vulva is obviously congested and swollen. The mucous inside the labia becomes reddish or hemoglobin with the estrus peak. The mucus turns large and thin, then the sexual desire tends to be vigorous. Subsequently, the appetite of sow rises, while looking sluggish. The vagina becomes light red, dry and slight wrinkled, the mucus turns from thin to thick, willing to accept humping (Figure 3-19). This is the end of estrus, which is the best time for insemination.

图 3-19　母猪的发情表现
Figure 3-19　Oestrus behavior of sows

2. 压背反射法

压背反射法是利用仿生学的手段鉴定母猪发情。在没有公猪在场的情况下，检查人员把手按压在母猪背部（图3-20）

1.2 Reflex of pressing back

Reflex of pressing back is a biomimetic method to identify estrus in sows. In the absence of boars, the inspectors press their hands on the back (Figure 3-20) or

或骑跨在母猪的背部，观察母猪的反应。也可利用试情公猪查情，检查人员用隔栏隔开公母猪，让公猪与母猪头对头，以便母猪能看到公猪并能嗅到公猪的气味，同时按压母猪背部（图3-21）。

straddle on the back of the sow to observe the response. We can also use a boar to test the sows. The inspectors separate the sows from the boars with fences, but let the sows see and smell the boars, and then check the reflex while pressing the back of sows (Figure 3-21).

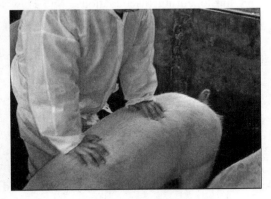

图3-20 猪的压背反射

Figure 3-20 Back pressure reflex of sows

母猪发情鉴定

Estrus identification of sows

图3-21 公猪试情

Figure 3-21 Boar to test the sows

如果按压背部母猪表现不安静，前后活动，表明尚在发情初期，或者已进入发情后期，不宜配种；如果按压后出现"静立反射"，即母猪不哼不叫，四肢叉开，呆立不动，弓腰，这是配种最适期。如果公猪在场，发情母猪表现为立耳、翘尾，主动接近公猪，压背时出现静立反射。

If sows are restless and moving around after pressing, it indicates that they are still in the early stage of estrus or the late stage of estrus. It is not suitable for mating at this time. If there is a "standing reflex" after pressing, that is, the sow does not hum or cry, but divides limbs, stands still, bows waist, this is the most suitable period for mating. If boars are present, estrus sows will also show erect ears and warped tails, while approaching boars actively, and standing reflex appears when pressing back.

Project Ⅲ Estrus Identification and Insemination

Estrus identification of sows is mainly based on external observation, combined with the reflex of pressing back method. In the production, the formula of "look firstly, listen secondly, count thirdly, press back fourthly and synthesize fifthly" have been summarized in the estrus identification of sows, namely: looking at vulva changes, behavior performance, feeding situation; listening to sow grunting; counting estrus cycle and estrus duration; doing back pressure test; conducting comprehensive analysis.

2 Insemination of sows

2.1 Preparations

2.1.1 Equipment

We should select different sizes of insemination tubes according to sow's condition (Figure 3-22). The insemination tube has spiral head or sponge head. Spiral head is usually made of rubber without side effects and is suitable for reserve sows. Sponge head is generally made of soft sponges and is suitable for multiparous sows. Special insemination tubes is used in deep insemination. Remove the tube from the sealed bag before insemination. Pay attention not to touch the first two-thirds of the tube. Apply lubricant on the fore end of the tube, and do not block the little hole at the top.

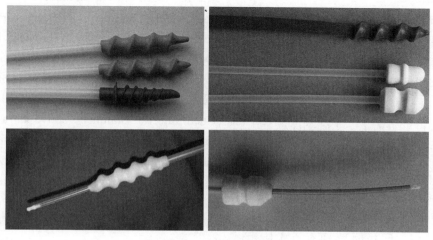

图 3-22 猪用输精管

Figure 3-22 Insemination tubes for sows

2. 精液

从保温箱取出输精瓶,轻轻颠倒几次。输精前,检查精子活率,精子活率低于0.6的精液不能使用。

3. 母猪

用0.1%的高锰酸钾溶液清洗消毒发情母猪外阴部。

(二)输精操作

1. 输精时间

一般在母猪发情后24~48h内配种容易受胎。老龄母猪发情时间较短,排卵时间会提前,应提前配种;青年母猪发情时间长,排卵期相应后移,宜晚配;中年母猪发情时间适中,应该在发情中期配种。

生产上,一般上午发现静立反射的母猪,下午输精一次,第二天下午进行第二次输精;下午发现静立反射的母猪,第二天上午输精一次,第三天上午再进行第二次输精。两次输精时间间隔至少8h。

2. 输精方法

母猪的阴道部和子宫颈结合处界限不明显,可直接将输精管插入阴道。输精时,将输精胶管涂以少量润滑剂,一只手将母猪阴唇分开,另一只手持输精管呈45°角斜上方插入母猪阴道内,避开尿道口后再平直前进。当感到向前推进有阻力时,说明海绵头已到达子宫颈外口,然后将输精管左右旋转推送3~5cm,当海绵头插入子宫颈管内后,子宫颈管受到刺激会收缩,使海绵头锁定在子宫颈管内。回拉时感到有一定阻力,连接输精瓶(袋),缓慢输入精液

2.1.2 Semen

Take out the semen from the incubator and gently turn it upside down several times. Before insemination, sperm viability should be examined. If sperm viability is below 0.6, the semen can not be used.

2.1.3 Sows

Wash and disinfect the vulva of sows with 0.1% potassium permanganate solution.

2.2 Insemination

2.2.1 Insemination time

Generally, sows are easy to conceive within 24-48 hours after estrus. Because the estrus time of old sow is short and the ovulation time will be ahead of time, so it should be mated in advance. On contrary, the estrus time of young sow is long and the ovulation period will be moved back, so it is suitable for late mating. The estrus time of middle-aged sow is moderate, so it should be mated in the middle of estrus.

In production, sows with standing reflex in the morning should be inseminated once in the afternoon, and re-inseminated in the afternoon of the next day. Those with standing reflex in the afternoon should be inseminated once in the next morning and repeat in the third day morning. The interval between two inseminations should be at least 8 hours.

2.2.2 Insemination methods

The boundary between vagina and cervix of sow is not obvious, and the semen injector can be inserted directly into vagina. During insemination, apply a small amount of lubricant to the semen injector, separate the labia of the sow with one hand, and insert the semen injector obliquely above at a 45-degree angle into the vagina with the other hand to avoid the urethral orifice, then move straight forward. When there is resistance to move forward, it means that the sponge head has reached the outer cervical orifice, and then the vas deferens are rotated and pushed forward for about 3-5 cm. When the sponge head is inserted into the cervical canal, the cervical canal will be stimu-

(图 3-23)。

在输精过程中,输精员同时按摩母猪阴户或大腿内侧,以刺激母猪的性兴奋,使其子宫收缩产生负压,促进精液吸收。输精时不要太快,一般需 3～10min 输完。输精完毕缓慢抽出输精管,并用手指按压母猪臀部使其安静片刻,以防精液倒流。

lated and contracted, so that the sponge head is locked in the cervical canal. When pulling back, you can feel a certain resistance, signaling the time for connecting the vase(bag) and slowly injecting semen(Figure 3-23).

In the process of insemination, the inseminator should massage the vulva or inner thigh of the sow to stimulate its sexual excitement. And this can make its uterus contract to produce negative pressure, and promote semen absorption. Insemination should not be too fast, it usually takes 3-10 minutes to complete. After insemination, vas deferens should be slowly drawn out, and press the buttocks of the sow to keep them stay for a moment, which could prevent the reflux of semen.

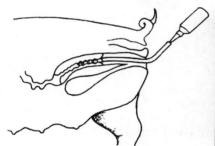

图 3-23　母猪的输精
Figure 3-23　Insemination of sows

母猪输精
Insemination of sows

任务 4　鸡的人工输精
Task 4　Insemination of Hens

任务描述 / Task Description

随着我国集约化养殖的快速发展,规模化鸡场普遍采用人工授精,提高了种蛋的受精率,降低了饲养成本。人工输精是一项认真、细致且具有技术性的工作,不同的输精部位和深度对受精率有何影响?

With the rapid development of intensive farm in China, artificial insemination is widely used in large-scale chicken farms, which improves the fertilization rate of eggs and reduces the feeding cost. Artificial insemination is a detailed, meticulous and highly technical work. What are the effects of different positions and depths of insemination on fertilization rate?

任务实施 / Task Implementation

一、准备工作

1. 母鸡的选择

输精母鸡应是营养中等、泄殖腔无炎症的母鸡。开始输精的最佳时间应为产蛋率达到70%的种鸡群。

2. 器具及用品准备

鸡用输精器（图 2-24）、原精液或稀释后的精液、酒精棉球等。

1 Preparations

1.1 Hens

Fertilized hens should be medium nourished and with no inflammation in the cloaca. The best time for insemination is when the laying rate reaches 70%.

1.2 Equipment

Prepare several inseminating syringes (Figure 3-24), original semen or diluted semen, alcohol cotton ball, etc.

图 3-24 鸡用输精器

Figure 3-24　Inseminating syringes for hens

二、输精要求

鸡适宜的输精间隔为 5~7d，每次输入原精液 0.025~0.03mL 或稀释精液 0.1mL，输入有效精子数至少为 5 000万个。输精时间应选择在大部分母鸡产蛋之后进行，最好在 16：00 左右。

三、输精操作

泄殖腔外翻法是输精最常使用的方法。翻肛员右手打开笼门，左手伸入笼内抓住母鸡双腿，把鸡的尾部拉出笼门口外，右手拇指与其他四指分开横跨于肛门两侧的柔软部分向下按压，当给母鸡腹部施加压力时，泄殖腔便可外翻，露出输卵管口（图 3-25）。此时，输精员手持输精枪对准输卵管开口中央，

2 Insemination requirements

The suitable interval of insemination for hens is 5-7 days. The insemination volume should be 0.025-0.03 mL original semen or 0.1 mL diluted semen. The number of effective sperm should be at least 50 million. The insemination time should be arranged after the laying time of most hens, preferably around 16：00.

3 Insemination

Everted cloaca method is the most commonly used method for hen insemination. With right hand opening the cage, the holder reaches into the cage with his left hand, grabs the two legs of the hen and pulls the tail out of the cage door. After fixing the hen, press the soft part of lower abdomen with the right hand. It is easier if we put the right thumb separated from the other four fingers across both sides of the anus. When pressure is applied to the hen's abdomen, the cloaca can

插入1~2cm注入精液。在输入精液的同时，翻肛员立即松手解除对母鸡腹部的压力，输卵管口便可缩回而将精液吸入。

be turned outwards, exposing the oviduct (Figure 3-25). At this time, the operator could insert into the oviduct at 1-2 cm and inject the semen. When the operator is injecting the semen, the holder should release the pressure on the abdomen of the hen, thus the oviduct outlet retracted and the semen could be absorbed.

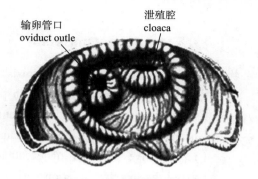

图 3-25 母鸡输精部位示意
Figure 3-25 Insemination position of hens

母鸡输精
Insemination of hens

四、注意事项

（1）输精前2~3h禁食禁水。

（2）捉取母鸡和输精动作要轻缓，尽量减少母鸡的恐惧感，防止引起鸡群骚动。

（3）输精时遇有硬壳蛋时动作要轻，将输精管偏向一侧缓缓插入输精。

（4）输精深度要适当，一般轻型蛋鸡采用浅阴道输精，即插入阴道1~2cm；中型蛋鸡或肉种鸡，应插入阴道2~3cm输精；母鸡产蛋率下降或精液品质较差时，插入阴道4~5cm输精。

（5）每只母鸡输一次应更换一支输精管，以防交叉感染。

（6）母鸡在产蛋期间，输卵管开口易翻出，每周重复输精一次，可保证较高的受精率。

4 Cautions

(1) Fasting and water deprivation for 2-3 hours before insemination.

(2) Grabbing hens and insemination should be done slightly, so as to minimize the fear and prevent the disturbance.

(3) If there is hard egg during insemination, the action should be slight, and syringe should be inserted into the oviduct slowly along one side.

(4) The depth should be appropriate. Generally, light laying hens use shallow vaginal insemination, i.e. inserting 1-2 cm. Medium laying hens or broiler breeders should insert 2-3 cm. When the laying rate of hens declines or the quality of semen is poor, insert 4-5 cm.

(5) To prevent cross infection, the syringe should not be reused.

(6) During laying period, the oviduct outlet can be everted easily, repeat insemination once a week can ensure a higher fertilization rate.

项目四 胚胎移植
Project Ⅳ Embryo Transfer

▲ 项目导学

自然条件下，牛、马等单胎动物通常一年产1胎，一生繁殖后代仅16只左右；猪也不过百头。利用胚胎移植可开发优良母畜的繁殖潜力，免去冗长的妊娠期，胚胎取出不久即可再次发情、配种和受孕，快速扩大良种畜群，因此掌握胚胎移植技术十分必要。

▲ Project Guidance

In nature, singleton animals such as cattle and horses usually produce one birth per year, only about 16 offsprings are born in a lifetime. And pigs have nearly 100 offsprings. Embryo transfer can develop the reproductive potential of excellent female animals, avoid the lengthy gestation period. And they could be in estrus, mating and conceived again soon after the embryo is taken out, so it is necessary to master the technology of embryo transfer.

◉ 学习目标

>>> 知识目标

- 掌握胚胎移植的概念。
- 理解胚胎移植的生理学基础和基本原则。
- 熟悉胚胎移植的技术流程。
- 了解体外受精、性别控制、胚胎分割、胚胎嵌合、细胞核移植等繁殖新技术。

>>> 技能目标

- 熟练掌握母畜的同期发情和超数排卵技术。
- 规范完成牛、羊的胚胎移植操作。

◉ Learning Objectives

>>> Knowledge Objectives

- To master the concept of embryo transfer.
- To understand the physiological basis and basic principles of embryo transfer.
- To be familiar with the technological process of embryo transfer.
- To know some new reproductive technologies such as in vitro fertilization, sex control, embryo splitting, embryo chimerism, nuclear transfer, etc.

>>> Skill Objectives

- To master the technology of estrus synchronization and superovulation of female animals.
- To complete the embryo transfer of cattle and sheep standardly.

一、胚胎移植的概念

胚胎移植俗称"借腹怀胎",是指将哺乳动物体内或体外生产的早期胚胎移植到同种的、生理状况相似的雌性动物生殖道内,使之继续发育成新个体。通常,将提供胚胎的个体称为供体,接受胚胎的个体称为受体。实际上,胚胎移植就是产生胚胎的供体和孕育胚胎的受体分工合作共同繁衍后代的过程,其中供体决定移植后代的遗传特性,受体只影响其体质发育。

二、胚胎移植的发展简况

1890年,英国剑桥大学的Walter Heape首次将纯种安哥拉兔的2枚胚胎移植到一只纯种比利时兔的输卵管内(用同种公兔交配后3h),结果生出了2只安哥拉仔兔和4只比利时仔兔,首次证实了胚胎移植技术的可行性。

家畜的胚胎移植开始于20世纪30年代。1934年首先在绵羊上获得成功,随后又相继在山羊(1949年)、猪(1951年)、牛(1951年)和马(1973年)上取得成功。1971年,世界上出现了第一个胚胎移植公司。1975年1月,国际胚胎移植学会成立大会在美国科罗拉多州丹佛召开。目前,美国、法国、德国、澳大利亚、加拿大、日本等国都已建立了牛胚胎移植的商业机构。

我国家畜胚胎移植的研究始于20世纪70年代,先后在兔(1973年)、绵羊(1974年)、牛(1977年)、山羊(1980

1 The concept of embryo transfer

Embryo transfer refers to the transfer of early embryos produced in vivo or in vitro from mammals into the reproductive tract of female animals with similar physiological status, so that they can continue to develop into new individuals. Usually, the one who provides the embryo is called the donor, and the one who receives the embryo is called the receptor. In fact, embryo transfer is a cooperational process between the donor and receptor. The donor determines the genetic characteristics of the transplanted offspring, and the receptor only affects their physical development.

2 Development of embryo transfer

In 1890, Walter Heape of Cambridge University first transplanted two embryos of pure Angolan rabbit into the oviduct of a pure Belgian rabbit(3 hours after mating with the same male rabbit). As a result, two Angolan rabbits and four Belgian rabbits were born, which proved the feasibility of embryo transfer technology for the first time.

Embryo transfer of livestock began in the 1930s. In 1934, it was first succeeded in sheep, and then succeeded in goats(1949), pigs(1951), cattle(1951) and horses(1973). In 1971, the first embryo transfer company appeared in the world. In January 1975, the founding conference of the International Society for Embryo Transfer was held in Denver, Colorado, USA. At present, the United States, France, Germany, Australia, Canada, Japan and other countries have established commercial institutions for cattle embryo transfer.

The research of livestock embryo transfer in China began in 1970s. It has succeeded in rabbits(1973), sheep(1974), cattle(1977), goats(1980), and horses

年)、马(1982年)等家畜上获得成功,90年代开始逐步在牛生产中推广应用。

三、胚胎移植的意义

1. 充分发挥优良母畜的繁殖潜力

通过胚胎移植,可使供体母畜省去冗长的妊娠期,从而缩短了繁殖周期。若再结合超数排卵技术,一次即可获得更多的优良胚胎,大大提高了繁殖效率,尤其是对牛、羊等单胎家畜。在自然繁殖状态下,一头母牛平均每年只能获得1头犊牛,利用胚胎移植技术,一头良种母牛一年能获得25～30头犊牛,最多可达50多头。

2. 代替种畜的引进

种畜引进价格高、运输不便、检疫和隔离程序复杂。胚胎移植不受时间和地点的限制,这样就可以通过胚胎的运输代替种畜的引进,节约购置和运输活畜的费用。此外,胚胎移植后代容易适应本地区的环境条件,也可以从养母得到一定的免疫力。

3. 保存品种资源

胚胎的冷冻保存,可以避免活畜保种因疾病、自然灾害而灭绝的风险,且成本低、易实施。胚胎移植不仅保存了珍贵的遗传资源,也为珍稀动物的繁衍和国际交流创造了机遇。

4. 克服不孕

有些优良母畜容易发生习惯性流产或难产,或者由于其他原因不宜进行妊娠过程的,可以采用胚胎移植使之正常繁殖后代。

(1982). And it has been gradually applied in cattle production since 1990s.

3 Importance of embryo transfer

3.1 Bringing into full play the reproductive potential of excellent female animals

By embryo transfer, donors can avoid long gestation period and shorten the reproductive cycle. Combined with superovulation, more excellent embryos can be obtained at one time, which greatly improves the reproductive efficiency, especially for singleton animals such as cattle and sheep. In natural, a cow can only give birth to a calf every year. By embryo transfer, a good cow can get 25-30 calves a year, up to 50 calves.

3.2 Replacing the introduction of breeding animals

The introducing of breeding animals is costly, transportation is inconvenient, procedures of quarantine and isolation are complex. Embryo transfer is not limited by time and place, so the introduction of breeding animals can be replaced by the transport of embryos, and the cost of purchasing and transporting live animals can be saved. In addition, the offsprings of embryo transfer are easy to adapt to the local environmental, and can also get certain immunity from fostress.

3.3 Preservation of variety resources

Embryo cryopreservation can avoid the risk of extinction due to diseases and natural disasters, and it is cheap and easy to implement. Embryo transfer not only preserves precious genetic resources, but also creates opportunities for the reproduction of rare animals and international exchanges.

3.4 Overcome infertility

Some excellent female animals are prone to habitual abortion or dystocia, some are not suitable to bear the pregnancy process for other reasons. By embryo transfer, they can reproduce normally.

5. 利于防疫

在养猪业中，为了培育无特定病原体（SPF）猪群，向封闭猪群引进新的个体时，为了控制疾病，往往采用胚胎移植技术代替剖腹取仔的办法。

6. 生物学的研究手段

胚胎移植是研究受精过程、胚胎学和遗传学等基础理论的有效方法之一，也是胚胎分割、胚胎嵌合、体外受精、性别鉴定、核移植、基因导入等其他胚胎工程技术实施的必不可少的环节。

四、胚胎移植的生理学基础与基本原则

（一）胚胎移植的生理学基础

1. 母畜发情后生殖器官的孕向发育

大多数自发性排卵的动物，发情后不论是否配种，或配种后是否受精，生殖器官都会发生一系列变化，如卵巢上出现黄体、子宫内膜增生、子宫腺体发育等，为早期胚胎的发育创造良好环境。

2. 早期胚胎的游离状态

胚胎在附植之前处于游离状态，尚未与子宫建立组织联系，营养主要来源于胚胎内的卵黄。这一特性是胚胎采集、保存、培养等操作的理论基础。

3. 子宫对早期胚胎的免疫耐受性

在妊娠期内，母体局部免疫逐渐发生变化，加之胚胎表面特殊免疫保护物质的存在，受体母畜对同种胚胎和胎膜组织一般不发生免疫排斥反应。

3.5 Benefiting epidemic prevention

In pig industry, in order to breed specific pathogen free (SPF) pigs and introduce new individuals to closed pig herds, embryo transfer is often used to replace caesarean section for disease control.

3.6 Research means of biology

Embryo transfer is one of the effective methods to study the basic theories of fertilization, embryology and genetics, and it is also an essential link in the implementation of other embryo engineering technologies, such as embryo splitting, embryo chimerism, in vitro fertilization, sex identification, nuclear transfer and gene introduction etc.

4 Physiological basis and basic principles of embryo transfer

4.1 Physiological basis of embryo transfer

4.1.1 Pregnancy development of female reproductive organs after estrus

Most spontaneously ovulated animals will undergo a series of changes in their reproductive organs, whether they are mating or not after estrus, fertilized or not after mating, such as corpus luteum on the ovary, endometrial hyperplasia and uterine glands development etc., which create a good environment for the development of early embryos.

4.1.2 Free state of early embryos

The embryo is in a free state before implantation, and it has not established a tissue connection with the uterus. Its nutrition mainly comes from yolk of embryo. This characteristic is the theoretical basis of embryo collection, preservation and culture etc.

4.1.3 Immunological tolerance of uterus to early embryos

During pregnancy, due to the gradual changes of maternal local immunity and the existence of special immune protective substances on the embryo surface, the receptor generally does not have immune rejection to the same embryo and fetal membranes.

4. 胚胎遗传物质的稳定性

胚胎的遗传信息在受精时就已确定,受体仅影响其体质发育,而不能改变其遗传特性。

(二) 胚胎移植的基本原则
1. 胚胎移植前后环境的一致性

(1) 分类学上的一致性。一般来讲,亲缘关系较远的物种,胚胎的生物学特性、发育条件、发育速度以及母体的子宫环境差异较大,胚胎与受体之间无法进行妊娠识别。因此,供体和受体一般要求为同种,但这并不排除异种间(在进化史上血缘关系较近、解剖和生理特点相似)胚胎移植的可能性。

(2) 生理学上的一致性。即供体和受体在发情时间上的同期性,一般要求相差不超过24h。

(3) 解剖部位的一致性。胚胎移植前后所处的空间环境要相似,即从供体输卵管内采集的胚胎应移植到受体的输卵管内,从供体子宫内采集的胚胎应该移植到受体的子宫内。

2. 胚胎的发育期限

从生理学角度讲,胚胎采集和移植的期限不应超过周期黄体的寿命,最迟要在黄体退化之前数日进行。因此,胚胎采集多在配种后3~8d进行,受体也应在相同时间内接受胚胎移植。

3. 胚胎的质量

在胚胎移植的过程中,胚胎不应受到任何不良因素的影响,移植的胚胎必须经过鉴定确认为发育正常的胚胎。

4.1.4 Stability of embryonic genetic material

The genetic information of the embryo has been determined at the time of fertilization. The receptor only affects its physical development, but can not change its genetic characteristics.

4.2 Basic principles of embryo transfer
4.2.1 Consistency of environments before and after embryo transfer

(1) Taxonomic consistency. The biological characteristics, developmental conditions, developmental speed of embryos and maternal uterine environment differ greatly among species with distant relationships, and pregnancy can not be identified between embryos and receptors. Therefore, donors and receptors should be homologous generally, but this does not exclude the possibility of embryo transfer between different species (closely related in evolutionary history, similar in anatomical and physiological characteristics).

(2) Physiological consistency. That is to say, the estrus is synchronous between donors and receptors, generally the difference should not exceed 24 hours.

(3) Consistency of anatomical location. The spatial environment should be similar before and after embryo transfer, that is, embryos collected from the donor's oviduct should be transferred to the receptor's oviduct, and embryos collected from the donor's uterus should be transferred to the receptor's uterus.

4.2.2 Development period of embryos

In the view of physiology, the period of embryo collection and transfer should not exceed the lifespan of the cycle of corpus luteum, several days before corpus luteum regression at latest. Therefore, embryo collection is usually carried out within 3-8 days after mating, and the receptor should receive embryos at the same time.

4.2.3 Quality of embryos

In the process of embryo transfer, embryos should not be affected by any harmful factors. The transferred embryos must be identified as normally developed.

4. 经济效益或科学价值

应用胚胎移植技术时，应考虑成本和最终收益。通常，供体胚胎应具有独特的经济价值，如生产性能优异或科研价值重大。

五、胚胎移植的操作程序

（一）供体和受体的选择

1. 供体的选择

供体应该是良种母畜，具有较高的育种价值和良好的繁殖性能，对超数排卵反应处理好。经产母畜应在生殖机能恢复正常后方可作为供体。

2. 受体的选择

受体可选用生产性能一般，但应具有良好的繁殖性能和健康状态。

（二）供体的超数排卵

在母畜发情周期的适当时期，利用外源促性腺激素进行处理，从而增加卵巢的生理活性，诱发多个卵泡同步发育成熟并排卵。

（三）供体的配种

为了获得较多发育正常的胚胎，对供体配种时应使用精子活率高、精子密度大的优质精液，适当增加输精次数，输精间隔为8～10h。

（四）胚胎的采集

胚胎的采集又称为冲胚，是利用冲胚液将早期胚胎从子宫或输卵管内冲出并收集的过程。胚胎采集分为手术法和非手术法两种，前者适用于各种家畜，后者仅适用于牛、马等大家畜，且只能

4.2.4 Economic or scientific value

When applying embryo transfer technology, the cost and benefit should be considered. In general, embryos should have unique economic value, such as excellent production performance or great scientific research value.

5 Operational procedures for embryo transfer

5.1 Selection of donors and recipients

5.1.1 Selection of donors

The donor should be an excellent female animal, with high breeding value and good reproductive performance. And it reacts well to superovulation. The multiparous dams should not be used as a donor until the reproductive function returns to normal.

5.1.2 Selection of recipients

Recipients with moderate production performance can be selected, but it should have good reproductive performance and good health.

5.2 Superovulation of donors

In the proper period of estrus cycle of female animals, exogenous gonadotrophin was used to treat them, which can increase the physiological activity of ovary, and induce multiple follicles to mature synchronously and ovulate.

5.3 Mating of donors

In order to obtain more normal embryos, excellent semen with high viability and high density should be used for donor mating. The number of inseminations should be increased appropriately and the interval between inseminations should be 8-10 hours.

5.4 Embryo collection

Embryo collection, also known as embryo flushing, is the process of using flushing fluid to rinse early embryos out of the uterus or oviduct and collect them. There are two methods for embryo collection: surgical method and non-surgical method. The former is suitable for all kinds of livestock, while the latter is

在胚胎进入子宫角以后进行。各种家畜的胚胎发育速度见表4-1。采集胚胎的时间一般在配种后3～8d，发育至4～8细胞或8细胞以上为宜，牛最好在配种后6～8d，此时胚胎发育至桑葚胚或者早期囊胚，便于非手术法采集和移植。

only used for cattle and horses etc, and it can only be carried out after the embryo enters the uterine horn. The embryo development speeds of various animals are shown in Table 4-1. The time of embryo collection is usually 3-8 days after mating, and it is suitable to develop to 4-8 cells or more. The best time for cattle is 6-8 days after mating, when the embryo develops to morula or early blastocyst, and it is convenient for non-surgical collection and transplantation.

表 4-1　各种家畜的胚胎发育速度（排卵后天数）
Table 4-1　Embryo development speeds of various animals (days after ovulation)

动物 Animal	2 细胞 2 cells	4 细胞 4 cells	8 细胞 8 cells	16 细胞 16 cells	进入子宫 Entering the uterus	囊胚形成 Blastocyst formation	附植 Implantation
牛 Cattle	1～1.5	2～3	3	3～4	4～5	7～8	22
绵羊 Sheep	1.5	2	2.5	3	3～4	6～7	15
猪 Pig	1～2	2～3	3～4	4～5	5～6	6	13
兔 Rabbit	1	1～1.5	1.5～2	2～3	2.5～3	3～4	

1. 手术法

手术法多用于羊、猪和兔等小动物。即通过外科手术在腹中线做一切口，轻轻拉出子宫角和输卵管，根据胚胎的发育阶段，选择不同的冲胚方法（图4-1）。

（1）输卵管采胚法。当胚胎处于输卵管时（排卵后1～3d），采用此方法。对于羊、兔，用带有磨钝针头的注射器刺入子宫角尖端，注入冲胚液，在输卵管伞部接取；对于猪，则可反向冲取。

5.4.1 Surgical method

It's mostly used in small animals such as sheep, pigs and rabbits. Through surgical incision in the ventrimeson, the uterine horn and oviduct are pulled out gently. We choose different methods of rinsing embryo according to the development stage of embryo (Figure 4-1).

（1）Embryo collection from oviduct.

This method is used when the embryo is in the oviduct (1-3 days after ovulation). For sheep and rabbits, a syringe with a blunt needle is inserted into the tip of the uterine horn, injecting flushing fluid and picking up the embryo at the oviduct umbrella. For pigs, flushing can be reversed.

（2）子宫角采胚法。当确认胚胎已进入子宫角内，可采用此法，即从子宫角基部注入冲胚液，在子宫角尖端接取，或者反向冲洗。

(2) Embryo collection from uterine horn.

This method is used when the embryo has entered the uterine horn. That is, injecting flushing fluid from the base of the uterine horn and picking up the embryo at the tip of uterine horn, or back washing.

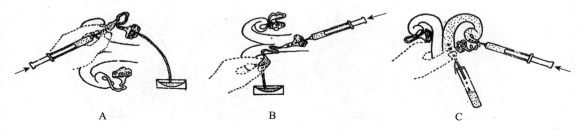

图 4-1　手术法冲胚示意

Figure 4-1　Surgical flushing embryos

A. 由子宫角向输卵管伞冲洗　B. 由输卵管伞向子宫角冲洗　C. 子宫角冲洗

A. flushing from uterine horn to oviduct umbrella　B. flushing from oviduct umbrella to uterine horn

C. flushing from uterine horn

2. 非手术法

对于牛、马等体型较大的动物，一般采用非手术法。非手术法回收胚胎都是在配种后 6～8d 进行，比手术法简单易行，对生殖器官的伤害较小。

（五）胚胎的检查

胚胎的检查是指将回收的胚胎置于体视显微镜下检查胚胎的数量和质量，并进行等级分类。不同发育阶段的正常牛胚胎见图 4-2。

生产中常用形态学方法进行胚胎的级别鉴定，通过观察胚胎的形态、卵裂球的大小与均匀度、色泽、细胞密度、与透明带间隙以及细胞变性等情况，将胚胎分为 A、B、C、D 四个等级，具体标准见表 4-2。

5.4.2　Non-surgical method

For large animals such as cattle and horses, non-surgical methods are generally used. Non-surgical embryo collection is carried out 6-8 days after mating, which is simpler and less harmful to reproductive organs than surgical method.

5.5　Embryo examination

Embryo examination is to check the quantity and quality of the recovered embryos under a stereomicroscope and grade them. The normal bovine embryos at different developmental stages are shown in Figure 4-2.

Embryos are classified into four grades: A, B, C and D by observing the morphology of embryos, the size and evenness of blastomere, color, cell density, clearance with zona pellucida and cell degeneration. The standards are shown in table 4-2.

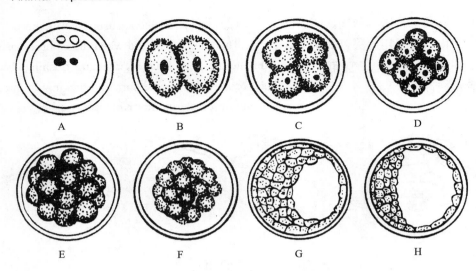

图 4-2 不同发育阶段的正常牛胚胎

Figure 4-2 Embryo development in the bovine

A. 1 细胞期（0～2d） B. 2 细胞期（1～3d） C. 4 细胞期（2～3d） D. 8 细胞期（3～5d）
E. 桑葚胚（6～7d） F. 紧实桑葚胚（6～8d） G. 早期囊胚（6～8d） H. 扩张囊胚（8～9d）
A. 1 cell（0-2 days） B. 2 cells（1-3 days） C. 4 cells（2-3 days） D. 8 cells（3-5 days）
E. morula（6-7 days） F. tight morula（6-8 days）
G. early blastocyst（6-8 days） H. expanded blastocyst（8-9 days）

表 4-2 胚胎分级标准

Table 4-2 Embryo grading standards

级别 Grade	标准 Standards
A	胚胎发育阶段与胚龄一致，形态完整，卵裂球轮廓清晰，大小均匀，结构紧凑，细胞密度大；色调和透明度适中，基本无游离细胞，变性细胞比例＜10% The embryo development stage is consistent with the embryonic age and the morphology is complete. The blastomere is clear in outline, uniform in size, compact in structure and dense in cell density. Tint and transparency is moderate, and it almost has no free cells, the percentage of degenerated cells is less than 10%.
B	胚胎发育阶段与胚龄基本一致，形态完整；卵裂球轮廓清晰，大小基本一致，细胞结合略显松散，密度较大；色调和透明度适中，变性细胞比例为 10%～20% The embryo development stage is basically aligned with the embryo age and the morphology is complete. The blastomere is clear in outline, basically same in size, slightly loose in structure and dense in cell density. Tint and transparency is moderate and the percentage of degenerated cells is 10%-20%.
C	胚胎发育阶段比正常迟缓 1～2d；卵裂球轮廓不清晰，大小不均匀；色调变暗，结构松散，游离的细胞较多，变性细胞达 30%～40% Embryo development is 1-2 days slower than normal. The blastomere is not clear in outline and uneven in size. Tint is darkening and structure is loose, it has more free cells and the percentage of degenerated cells is up to 30%-40%.
D	胚胎发育迟缓 2d 以上，细胞团破碎，变性细胞比例超过 50% Embryo development is 2 days slower than normal, cell mass is broken and the percentage of degenerated cells is more than 50%.

（六）胚胎的保存

胚胎的保存是指将胚胎在体内或体外正常发育温度下，暂时贮存起来而不使其活力丧失；或将其保存于低温或超低温条件下，使细胞新陈代谢和分裂速度减慢或停止，一旦恢复正常发育温度，又能继续发育。目前，哺乳动物胚胎的保存方法较多，主要包括常温保存、低温保存和超低温冷冻保存等。

1. 常温保存

将胚胎置于15～25℃的培养液中保存，此温度下胚胎能保存24h，仅用于短暂的保存和运输。

2. 低温保存

将胚胎置于0～10℃保存，此时，胚胎卵裂暂停，代谢减慢。不同动物的胚胎对温度的反应不同，几种哺乳动物的适宜保存温度分别为：小鼠5～10℃、家兔10℃、绵羊10℃、山羊5～10℃、牛0～6℃、猪15～20℃为宜。

3. 超低温冷冻保存

将胚胎置于超低温环境中（－196℃）保存，其新陈代谢及发育暂时停止，可对胚胎进行长期保存。常用的方法有逐步降温法、快速冷冻法及玻璃化冷冻法。快速冷冻法需使用专门的冷冻仪，适合大规模胚胎生产使用。玻璃化冷冻法是将胚胎放入含高浓度抗冻剂的冷冻保存液中，通过快速降温使胚胎形成玻璃化状态，此法操作简便快速，但必须严格控制每一步操作环节。

（七）胚胎的移植

同采集胚胎一样，胚胎的移植也有手术法和非手术法两种，前者适于羊、

5.6 Embryo preservation

Embryo preservation is the temporary storage of embryos at normal developmental temperature in vivo or in vitro without loss of viability, or the storage of embryos at low or ultra-low temperatures, which slows down or stops the metabolism and division of cells. And the embryos can continue to develop once the normal developmental temperature is restored. At present, there are many ways to preserve mammalian embryos, including normal-temperature preservation, low-temperature preservation and cryopreservation.

5.6.1 Normal-temperature preservation

Embryos are preserved in nutrient solution at 15-25℃. At this temperature, the embryos could be preserved for 24 hours, which is only for short-term storage and transportation.

5.6.2 Low-temperature preservation

Embryos are preserved at 0-10℃. At this time, the cleavage of the embryo pauses and the metabolism slows down. Different animals have different responses to temperature variations. The suitable temperature is 5-10℃ in mice, 10℃ in rabbits, 10℃ in sheep, 5-10℃ in goats, 0-6℃ in cattle and 15-20℃ in pig.

5.6.3 Cryopreservation

Embryos can be preserved for a long time by putting them in ultra-low temperature(－196℃), and their metabolism and development are temporarily stopped. There are three methods: stepwise cooling, rapid freezing and vitrification. Rapid freezing needs special embryo freezer, which is suitable for large-scale embryo production. Vitrification is to put embryos into the cryopreservation solution with high concentration of antifreeze and make embryos vitrified by rapid cooling. This method is simple and fast, but every step of operation must be strictly controlled.

5.7 Embryo transfer

Like collecting embryos, there are two methods for embryo transfer: surgical method and non-surgical meth-

猪、兔等体型较小动物，后者仅适于牛、马等大家畜。在胚胎移植前，应检查受体母畜排卵一侧卵巢上的黄体数量及发育情况，只有黄体发育良好的受体才能用于胚胎移植。

od. The former is suitable for small animals such as sheep, pigs and rabbits, and the latter is only suitable for large animals such as cattle and horses. Before embryo transfer, the number and development of corpus luteum on the ovary with ovulation of the recipient should be checked. Only the recipients with well-developed corpus luteum can be selected for embryo transfer.

六、胚胎移植存在的主要问题

1. 超数排卵效果不稳定

超数排卵并不能每次都达到预期效果，不同批次的药品存在效价差异，不同的个体和年龄对超数排卵处理的反应差异很大，造成排卵率极不稳定。

2. 胚胎回收率较低

家畜在超数排卵处理后，排卵数过多往往会降低胚胎的回收率，其原因可能是由于卵巢在外源激素的作用下，体积增大，使输卵管伞难以完全包被卵巢，造成一些卵子丢失。一般情况下，胚胎的回收率在50%~80%。

6 Main problems in embryo transfer

6.1 The effect of superovulation is unstable

Superovulation can not achieve the desired effect every time. There are differences in the potency of drugs in different batches, and the response of different individuals and ages to superovulation varies greatly, so the ovulation rate is extremely unstable.

6.2 The recovery of embryos is low

After superovulation, excessive ovulation often reduces the recovery rate of embryos. The ovaries are enlarged with exogenous hormones, which makes it difficult for the oviduct umbrella to completely encapsulate the ovary, resulting in the loss of some ovums. Generally, the recovery rate of embryos is between 50% and 80%.

任务1 羊的胚胎移植
Task 1　Embryo Transfer in Sheep

任务描述

羊肉是集营养和保健于一体的肉食品，越来越受到消费者的青睐。目前，肉羊的供种体系不健全，种羊引进价格昂贵，在一定程度上制约了肉羊产业的快速发展。随着胚胎移植技术的大力推广，它在肉羊的纯种繁育、快速扩群及性能提高等方面具有重要作用。如何进行羊的胚胎移植？

Task Description

Mutton is fine meat for its nutrition and health-preserving properties, and enjoying growing popularity with consumers. At present, the supply system of breeding sheep is not perfect, and the introduction price of breeding sheep is expensive, which restricts the rapid development of sheep industry to some extent. With the extension of embryo transfer, it plays an important role in sheep breeding, rapid expansion and performance improvement. How to carry out embryo transfer in sheep?

任务实施

一、准备工作

（1）动物。供体母羊、受体母羊。

（2）仪器。超净工作台、恒温水浴锅、体视显微镜、内窥镜、CO_2培养箱、手术台等。

（3）材料。常规手术器械、平皿、量筒、移液器、细管（1/4）、注射器（20 G）、0.5%利多卡因、孕酮阴道栓（CIDR）、冲胚液、FSH、LH、PMSG、HCG、氯前列烯醇、青霉素、链霉素、蒸馏水、水温计等。

二、操作方法

（一）供、受体羊的选择

供体羊应健康无病，生产性能高，无遗传缺陷疾病。受体羊应选1～3岁的杂交一代或当地母羊，胎次为1～3胎，体格接近供体羊，健康无病，无繁殖障碍。

（二）同期发情与超数排卵

在供体羊发情周期第12天或第13天开始肌内注射FSH（150～300IU），以日递减剂量连续注射3d，每次间隔12h，在第5次注射FSH同时肌内注射$PGF_{2\alpha}$ 2mg。

受体羊在胚胎移植前7d注射$PGF_{2\alpha}$ 2mg，移植前6d注射FSH 50IU，每天试情并记录发情时间。

（三）胚胎的收集

羊的胚胎收集一般采用手术法。术前采用腹腔镜检查供体羊的卵巢和子宫，弃除不适宜手术的羊。

Task Implementation

1 Preparation

(1) Animals: donor and recipient ewes.

(2) Instruments: ultra-clean worktable, thermostatic water bath, stereo microscope, endoscope, CO_2 incubator, operating table.

(3) Materials: conventional surgical instruments, plate, graduated cylinder, transferpettor, tubule (1/4), injector (20G), 0.5% lidocaine, CIDR, flushing fluid, FSH, LH, PMSG, HCG, chloroprostenol, penicillin, streptomycin, distilled water, water thermometer.

2 Operation

2.1 Selection of donors and recipients

Donors should be healthy, with high production performance and no genetic defects. Recipients should be 1-3 years old hybrid or local ewes, with 1-3 calving times, close to donors in size, healthy and has no reproductive obstacles.

2.2 Estrus synchronization and superovulation

FSH (150-300 IU) is injected intramuscularly on the 12th or 13th day of the estrous cycle of the donor sheep. FSH is injected continuously for 3 days, then the dose decreases day by day, 12 hours interval each time, and $PGF_{2\alpha}$ is injected intramuscularly with a dose of 2mg at the same time of the 5th injection of FSH.

The recipient sheep are injected with 2mg $PGF_{2\alpha}$ on the 7th day and 50 IU FSH on the 6th day before embryo transfer. We test the oestrus every day and record the time of oestrus.

2.3 Embryo collection

Embryo collection in sheep usually uses surgical method. The ovaries and uterus of the donor sheep are examined by laparoscopy before operation, and the un-

1. 输卵管采胚

采集胚胎的时间一般在配种后 2~3d，从输卵管伞部插入细管，从宫管结合部向输卵管方向注入 5~10mL 冲胚液。

2. 子宫角采胚

在配种后 6~7d，胚胎已进入子宫角内部。用止血钳夹住子宫角基部，在子宫角基部插入细管，从子宫角尖端注入 50~60mL 冲胚液。

（四）胚胎的检查

将回收液置于无菌室静置 10~20min，吸出上清液，在体视显微镜下观察胚胎的形态和发育情况，收集外形整齐、大小一致、卵裂球分裂均匀、外膜完整的胚胎备用。

（五）胚胎的移植

1. 手术法

（1）术前 24h 禁食。

（2）剃毛。

（3）保定、麻醉。

（4）消毒。

（5）手术。在一侧乳房基部稍前方于腹中线旁 2~3cm 处，由后向前平行切开 5~8cm，寻找并显露子宫、输卵管及卵巢，观察并记录两侧卵巢表面的黄体大小和数量。

（6）移胚（图 4-3）。用钝形针头在有黄体一侧子宫角上部 1/3 处扎一小孔，将装有胚胎的移植器沿针孔插入子宫腔内约 1cm，注入胚胎，最后缝合切口。

（7）术后护理和妊娠诊断。为了防

suitable sheep were discarded.

2.3.1 Embryo collection from oviduct

The time of embryo collection is usually 2-3 days after mating. The tubule is inserted from oviduct umbrella, and 5-10 mL flushing fluid is injected from the uterotubal junction to the oviduct.

2.3.2 Embryo collection from uterine horn

The embryo has entered the uterine horn in 6-7 days after mating. The base of uterine horn is clamped with forceps hemostatic and inserted a tubule, 50-60 mL flushing fluid is injected from the tip of uterine horn.

2.4 Embryo examination

The recovered solution is placed in sterile room for 10-20 minutes, and the supernatant is sucked out. The morphology and development of embryos are observed under the stereomicroscope. The embryos with neat shape, uniform size, even blastomere and complete membrane are collected for reserve.

2.5 Embryo transfer

2.5.1 Surgical methods

(1) Fasting 24 hours before operation.

(2) Shaving.

(3) Fixing and anesthesia.

(4) Disinfection.

(5) Operation. At the slightly-frontward base of one side of the breast, beside the ventrimeson at 2-3cm, the uterus, oviducts and ovaries are found and exposed by cutting 5-8 cm parallel from the back to the front. The size and quantity of corpus luteum on the surface of the ovaries on both sides are observed and recorded.

(6) Embryo transfer (Figure 4-3). With a blunt needle, a small hole is pierced in the upper 1/3 of the uterine horn on the side of corpus luteum. The implanter with embryo is inserted into the uterine cavity with a depth about 1 centimeter, and then the embryo is injected into the uterus. Finally, the incision is sutured.

(7) Postoperative nursing and pregnancy diagno-

止术后引发感染，需连续注射抗生素3～7d，最好单圈饲养，做好保胎工作。

sis. In order to prevent infection after operation, antibiotics should be injected continuously for 3-7 days. It is better to raise the recipient sheep in a separate enclosure for tocolysis.

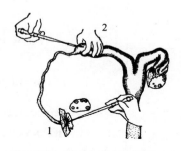

图 4-3 羊的手术法移植胚胎

Figure 4-3 Surgical transfer of embryos in sheep

1. 将胚胎移植于输卵管　2. 将胚胎移植于子宫角

1. embryo transfer to oviduct　2. embryo transfer to uterine horn

2. 腹腔内窥镜移植

对羊进行全身麻醉，在乳房前方腹中线两侧各切开一个长1 cm的小口，一侧插入打孔器和腹腔镜，另一侧插入宫颈钳。观察卵巢上黄体的情况，选择黄体发育良好一侧的输卵管或子宫角移植。该法对羊的损害小，整个移植过程仅需2～3 min，目前已成为羊胚胎移植的一种主要方法。

2.5.2　Endoscope transfer

Perform general anesthesia on the sheep, and cut a small opening of 1 cm in length on each side of the midline of the abdomen in front of the breast. A perforator and endoscope are inserted on one side, and cervical forcep is inserted on the other side. Observe the corpus luteum on the ovary and transfer the embryos into oviducts or uterine horn on the side with well-developed corpus luteum. This method has little damage to sheep and only takes 2-3 minutes for the whole process of embryo transfer. It has become the main method of embryo transfer in sheep.

任务2　牛的非手术法胚胎移植
Task 2　Non-surgical Embryo Transfer in Cattle

任务描述 / Task Description

某农业院校举办畜牧兽医人才招聘会，吸引了65家用人单位参加，其中4家公司招聘胚胎移植技术人员9人。这些公司主要致力于优质高产奶牛的推

An Agricultural University held a job fair for animal husbandry and veterinary talent, which attracted 65 companies, and 4 of them recruited 9 embryo transfer technicians. These companies are mainly committed to

广，通过胚胎移植技术进行品种改良和快速扩群。目前，奶牛的胚胎移植主要采用哪种方法？如何操作？

the promotion of high-quality and high-yield cows, through embryo transfer for breed improvement and rapid expansion. At present, which method is mainly used for embryo transfer of cows? How to operate?

任务实施

一、准备工作

（1）动物。供体母牛、受体母牛。

（2）仪器。超净工作台、恒温水浴锅、干燥箱、液氮罐、体视显微镜、CO_2 培养箱。

（3）材料。阴道开腔器、二路式或三路式冲胚管（图4-4）、平皿、量筒、移液器、细管（1/4）、注射器（20 G）、静松灵、2%利多卡因、冲胚液、犊牛血清、PMSG、FSH、LH、$PGF_{2\alpha}$、青霉素、链霉素、水温计等。

Task Implementation

1 Preparation

(1) Animals: donor cows and recipient cows.

(2) Instruments: ultra-clean worktable, thermostatic water bath, drying closet, nitrogen canister, stereo microscope and CO_2 incubator.

(3) Materials: vaginal opener, two-way or three-way flushing tube (Figure 4-4), plate, graduated cylinder, transferpettor, tubule (1/4), injector (20G), xylidinothiazoline, 2% lidocaine, flushing fluid, PMSG, FSH, LH, $PGF_{2\alpha}$, penicillin, streptomycin, water thermometer, etc.

图 4-4　冲胚管

Figure 4-4　Flushing tube

二、操作方法

（一）供、受体牛的选择

供体牛应健康无病，种用价值高，超排效果好；受体牛可选用非良种个体，但要求体型较大、膘情中等、繁殖性能良好。

（二）供体牛的超数排卵与配种

在发情周期第 11 天肌内注射 PMSG

2 Operation

2.1 Selection of donors and recipients

Donors should be healthy, with high breeding value and good superovulation effect. Recipients can be non-breeding individuals, but with large size, moderate fat and good reproductive performance.

2.2 Superovulation and mating of donors

PMSG (3 000-4 000 IU) is injected intramuscularly on

3 000～4 000IU，或在发情周期第11～14天连续4d肌内注射FSH，每天2次，间隔12 h，总剂量400 IU。在发情周期第13天肌内注射PGF$_{2\alpha}$ 20～30 mg，促使黄体退化。供体牛在发情周期第15天发情，发情后8～12 h首次输精，同时注射LH 160 IU，一般输精3次，间隔8 h。

（三）胚胎收集

母牛多采用非手术法收集胚胎（图4-5）。冲胚前禁水、禁食10～24 h，将供体牛保定于保定栏内，将尾巴拉向一侧，清除直肠内的宿粪，清洗阴户周围。

1. 麻醉

冲胚前10 min，剪去荐椎和第一尾椎结合处或第一尾椎和第二尾椎结合处的被毛，先用酒精消毒，然后肌内注射静松灵3～5 mL，再用2%利多卡因4 mL进行尾椎硬膜外麻醉。

the 11th day of the estrus cycle, or FSH is injected intramuscularly between the 11th and 14th day of estrus cycle for 4 consecutive days, twice a day, at 12 hours intervals, with a total dose of 400 IU. PGF$_{2\alpha}$ (20-30 mg) is injected intramuscularly on the 13th day of the estrus cycle to induce corpus luteum degeneration. The donor cattle is estrus on the 15th day of estrus cycle, and is inseminated 8-12 hours after estrus for the first time. At the same time, LH (160 IU) is injected. Insemination is performed three times at 8 hours intervals.

2.3 Embryo collection

Non-surgical method is mostly used to collect embryos in cows (Figure 4-5). Water and food are forbidden for 10-24 hours before embryo flushing. The donor cow is fixed. The tail is pulled to one side to remove the feces in the rectum and wash around the vulva.

2.3.1 Anesthesia

Ten minutes before embryo flushing, clothing hair at the junction of sacral vertebra and first coccygeal vertebra or the junction of first coccygeal vertebra and second vertebra is snipped. Disinfecting with alcohol, xylidinothiazoline is injected intramuscularly for 3-5 mL, then 4 mL 2% lidocaine is used for coccygeal vertebra epidural anesthesia.

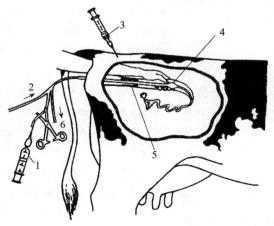

图4-5 牛的非手术法冲洗胚胎
Figure 4-5 Non-surgical embryo flushing in cows
1. 注气口 2. 冲胚液入口 3. 麻醉剂 4. 气囊 5. 子宫颈 6. 回收液出口
1. air injection port 2. import of flushing fluid 3. anesthetics 4. air sac 5. cervix 6. export of recycled liquid

2. 冲胚

组装冲胚装置，准备 1 000 mL 冲胚液加温至 38℃备用。对于青年母牛或子宫颈通过困难的供体牛，先使用开膣器扩张子宫颈口，然后将带有钢芯的冲胚管插入一侧子宫角大弯处，充气 15~20 mL，气囊胀大，使冲胚管固定于子宫基部，以免冲胚液流入子宫体并沿子宫颈口流失。抽出钢芯，用注射器吸取冲胚液 40~60 mL 注入子宫角，回收冲胚液，反复冲洗。用同样方法冲洗另一侧子宫角。冲洗完毕，向子宫内注入抗菌药物。

（四）胚胎检查
同羊的胚胎检查。

（五）胚胎移植

1. 受体牛的筛选

受体牛在发情后 6~8 d 均可进行移植。移植前，检查两侧卵巢的黄体发育情况，只有黄体发育合格者才能用于移植，并在臀部标记黄体所在一侧。

2. 麻醉

将受体牛保定，在第 1~2 尾椎间进行硬膜外麻醉，清除宿粪，冲洗外阴部，并用酒精消毒。

3. 移胚（图 4-6）

对照受体发情记录，选择适宜发育阶段和级别的胚胎，封装于 0.25 mL 塑料细管内，把装有胚胎的细管装入预温的移植枪管内，金属内芯轻轻插入细管的棉塞端内，套上灭菌的软外套。

2.3.2 Embryo flushing

Assemble the embryo flushing device and heating 1 000 mL flushing fluid to 38℃ for use. For young cows or donors whose cervix is difficult to pass through, first use the opener to dilate the cervix. Then the flushing tube with steel core is inserted into the greater curvature of uterine horn on one side. Inflate 15-20 mL, expand the air sac, and fix the embryo flushing tube at the base of the uterus to prevent flushing embryo fluid from flowing into the uterus and losing along the cervix. Withdraw the steel core, use a syringe to draw 40-60 mL of the embryo flushing fluid into the uterine horns, recover the embryo flushing fluid, and rinse repeatedly. The other uterine horn is rinsed in the same way. After rinsing, antibio-tics are injected into the uterus.

2.4 Embryo examination

The embryo examination is the same as that of sheep.

2.5 Embryo transfer

2.5.1 Screening of recipients

The recipients can be transplanted 6-8 days after estrus. Before transplantation, check the development of the corpus luteum on both ovaries. Only those cattle with qualified corpus luteum could be used for transplantation, and the side of the qualified corpus luteum should be marked on the buttocks.

2.5.2 Anesthesia

The recipient cow is fixed and anesthetized epidurally between the 1st and 2nd sacral vertebra. Feces are removed, vulva is washed, and disinfected with alcohol.

2.5.3 Embryo transfer (Figure 4-6)

According to the estrus records of recipients, embryos of suitable development stages and grades are selected and packed in 0.25 mL plastic tube. The tube is then put into the preheated transplantation pipe. The metal inner core is gently inserted into the cotton plug end of the tube, next the sterilized soft coat is put on.

在胚胎移植时，用直肠把握法将移植枪通过子宫颈插入黄体侧子宫角大弯处，缓慢推出胚胎，缓慢、旋转抽出移植枪，轻轻按摩子宫角 3～4 次。

During embryo transfer, the transfer gun through the cervix is inserted into the large curve of uterine horn on the corpus luteum side by rectal grasp, and the embryo is pushed out slowly. The gun is drawn out slowly and rotationally, and the uterine horn is massaged gently for 3-4 times.

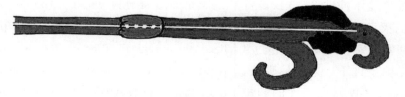

图 4-6　牛子宫角移植胚胎

Figure 4-6　Embryo transfer from uterine horn in cows

拓展知识 / Knowledge Expansion

一、体外受精

1　In vitro fertilization（IVF）

体外受精是动物胚胎生物工程的重要研究内容之一，是指哺乳动物的精子和卵子在体外人工控制的环境中完成受精过程的技术。早在 1878 年，德国科学家 Sckenk 就开始进行哺乳动物卵子体外受精尝试。1951 年 Chang 和 Austin 同时发现精子获能现象之后，哺乳动物体外受精的研究得到了飞速发展。1959 年 Chang 首次获得"试管兔"，为哺乳动物体外受精工程奠定了基础。到目前为止，已先后在家兔（1959 年）、小鼠（1968 年）、大鼠（1974 年）、婴儿（1978 年）、牛（1982 年）、山羊（1985 年）、绵羊（1985 年）和猪（1986 年）等动物获得了成功。

应用体外受精技术可获得大量胚胎，使胚胎生产"工厂化"，为胚胎移植及相应胚胎工程提供胚胎来源，不仅在畜牧业上具有广阔的应用前景，而且

In vitro fertilization is one of the important research contents of animal embryo bioengineering. It refers to the technology of mammalian sperm and ovum completing fertilization process in an artificial controlled environment in vitro. As early as 1878, Sckenk, a German scientist, attempted in vitro fertilization of ovum. After Chang and Austin discovered sperm capacitation in 1951, in vitro fertilization of mammals has developed rapidly. In 1959, Chang first obtained the "test-tube rabbit", which laid the foundation for mammalian in vitro fertilization. So far, it is successful in rabbits (1959), mice (1968), rats (1974), infants (1978), cattle (1982), goats (1985), sheep (1985) and pigs (1986) etc.

A large number of embryos can be obtained by in vitro fertilization, which makes mass embryo production possible and provides embryos for embryo transfer and embryo engineering. It not only has broad application pros-

在医学上可以治疗不孕症，同时对于丰富受精生物学的基础理论也具有重大意义。

（一）卵母细胞的采集

1. 离体卵母细胞采集

母畜被屠宰后，30 min 内无菌采集卵巢，用生理盐水冲洗 2～3 次，置于 30～35℃ 的灭菌生理盐水或 PBS 液中，尽快运回实验室，时间以不超过 4 h 为宜。在无菌条件下，采用抽吸法、切割法或剥离法采集卵巢表面的卵母细胞。

2. 活体卵母细胞采集

借助超声波探测仪或腹腔镜，经阴道穿刺，直接从活体动物的卵巢中吸取卵母细胞。牛常用超声波探测仪辅助取卵，操作者将探头插入阴道穹隆部，一只手伸入直肠把握卵巢，紧贴在探头所在的部位，借助 B 超图像引导，另一只手持吸卵针经阴道壁穿刺吸取卵母细胞。绵羊和猪等小动物常用腹腔镜取卵。

（二）卵母细胞的体外成熟

从卵巢上采集的卵母细胞尚未成熟，需要进一步培养才能与精子受精。用于卵母细胞体外成熟的培养液有多种，其中应用最广泛的是 TCM-199。研究发现，在培养液中加入 cAMP、生殖激素（如 FSH、17β-雌二醇、HCG 等）、血清、生长因子等，可促进卵母细胞成熟，提高培养效果。卵母细胞一般在 39℃、5% CO_2、饱和湿度条件下需培养 20～30h。卵丘细胞充分扩张，第一极体释放，从形态学上即可认为卵母细胞达到成熟。

pects in animal husbandry, but also can be used in clinical treatment of infertility, and has great significance for enriching the basic theory of fertilization biology.

1.1 Oocyte collection

1.1.1 In vitro oocyte collection

After slaughter, the ovaries of the female animals are collected sterilly within 30 minutes, washed with physiological saline for 2-3 times, and placed in sterilized physiological saline or PBS solution at 30-35℃. It was returned to laboratory as soon as possible for no more than 4 hours. Under sterile conditions, oocytes on ovarian surface are collected by suction, cutting or peeling.

1.1.2 In vivo oocyte collection

With the help of ultrasound detector or laparoscope, oocytes are directly extracted from the ovaries of living animals by vaginal puncture. Ultrasound detectors are often used to assist in oocyte retrieval in cattle. The operator inserts the probe into the vaginal fornix, one hand reaches into the rectum to grasp the ovary, and the other holds the oocyte suction needle through the vaginal wall to puncture and suck the oocytes. In small animals such as sheep and pigs we usually use laparoscopy.

1.2 In vitro maturation of oocytes

Oocytes collected from ovaries are immature and need further culture to fertilize with sperm. There are many kinds of nutrient solution for oocyte maturation in vitro, among which TCM-199 is the most widely used. It was found that the addition of cAMP, reproductive hormones (such as FSH, 17β-estradiol, HCG, etc.), serum and growth factors in the nutrient solution could promote oocyte maturation and improve the culture effects. Oocytes usually need to be cultured for 20-30 hours at 39℃, 5% CO_2 and saturated humidity. The cumulus cells are fully expanded and the first polar body is released. Morphologically, the oocytes can reach maturation.

(三) 精子体外获能

目前，精子获能的处理方法主要有肝素处理法、钙离子载体法和高渗溶液处理法。肝素是一种高度硫酸化的氨基多糖类化合物，与精子结合后，引起 Ca^{2+} 进入精子细胞内部而导致精子获能。钙离子载体能直接诱发 Ca^{2+} 进入精子细胞内部，提高细胞内的 Ca^{2+} 浓度，从而导致精子获能。精子表面含有许多膜蛋白，即所谓的"去能因子"。当用高渗溶液处理精子时，可促使膜蛋白脱落而使精子获能。

(四) 体外受精

将获能精子与成熟卵母细胞共同培养完成受精过程。体外受精的培养系统主要包括微滴法和四孔培养板法两类。微滴法是一种应用最广的培养系统，即在培养皿中将受精液做成 20～40 μL 的微滴，上覆石蜡油，每滴放入成熟卵母细胞 10～20 枚及获能精子（1.0～1.5）$\times 10^6$ 个/mL，孵育 6～24 h。如出现精子穿入卵内，头部膨大，第二极体排出，原核形成和正常卵裂等，即可确定为受精。

(五) 早期胚胎的体外培养

精子和卵子受精后，受精卵需移入胚胎培养液中继续培养至致密桑葚胚或早期囊胚阶段。许多动物早期胚胎均存在"体外发育阻滞"现象，即胚胎发育到一定时期会受到不同程度的阻滞，牛、绵羊为 8～16 细胞、山羊为 2 细胞、猪为 4 细胞。研究发现，在 38～39℃、

1.3 In vitro capacitation of sperm

At present, the main methods of sperm capacitation are heparin treatment, calcium ionophore and hyperosmotic solution treatment. Heparin is a highly sulfating amino-polysaccharide compound, which binds to sperm and causes Ca^{2+} to enter the sperm cells, resulting in sperm capacitation. Calcium ionophore can directly induce Ca^{2+} to enter sperm cells and increase the concentration of Ca^{2+} in sperm cells, thus leading to sperm capacitation. Membrane proteins on the sperm surface are the so-called "decapacitation factor". When sperm were treated with hyperosmotic solution, membrane proteins could be caducous and sperm could be capacitated.

1.4 In vitro fertilization

Fertilization was completed by co-culture of capacitated sperm and mature oocyte. The culture systems of in vitro fertilization mainly include microdrop sytem and four-hole plate system. The microdrop system is the most widely used culture system. Fertilizer fluid is made into 20-40 μL microdroplets in a culture dish and covered with liquid paraffin. Each drop is placed into 10-20 mature oocytes and capacitated sperm $(1.0$-$1.5) \times 10^6$/mL, then incubated for 6-24 hours. If sperm penetrates the ovum, head expands, the second polar body is discharged, pronucleus forms and cleavage occurs, fertilization can be concluded as completed.

1.5 In vitro culture of early embryos

After fertilization, the fertilized ovum needs to be transferred into the embryo nutrient solution to develop into tight morula or early blastocyst stages. Many animal embryos have "developmental block *in vitro*", that is, embryonic development will be stunted in a certain period, cattle and sheep are at 8-16 cells stage, goats are at 2 cells stage, pigs are at 4 cells stage. It was

5%CO_2、饱和湿度、碳酸氢钠或TCM-199培养条件下,采用与其他细胞共同培养的方法可促进早期胚胎的体外发育。这类细胞的类型很多,如卵丘颗粒细胞、成纤维细胞、滋养层细胞、黄体细胞等,但仍存在着囊胚发育率低的问题,有待于进一步改善培养条件。

二、 克隆技术

在生物学中,克隆是指由一个细胞或个体以无性繁殖的方式产生遗传物质完全相同的一群细胞或一群个体。在动物繁殖学中,它是指不通过精子和卵子的受精过程而产生遗传物质完全相同的新个体的一种胚胎生物技术。

(一) 胚胎分割

胚胎分割是运用显微技术人工制造同卵双胎或同卵多胎的技术,是扩大良种胚胎来源的一条重要途径,其理论依据是早期胚胎的每一个卵裂球都具有独立发育成新个体的全能性。

不同阶段的胚胎,分割方法略有差异。桑葚胚之前的胚胎,因卵裂球较大,直接分割对卵裂球损伤较大,常采用卵裂球分离方法;桑葚胚或囊胚,卵裂球结合紧密,细胞界限逐渐消失,多采用胚胎切割方法。在进行囊胚分割时,要注意将内细胞团等分。胚胎分割技术已在多种动物取得成功,但仍存在很多问题,需作深入研究。例如,初生重小,遗传不完全一致,异常与畸形等。

found that co-culture with other cells could promote the development of early embryos in vitro under the conditions of 38-39℃, 5% CO_2, with saturated humidity, sodium bicarbonate or TCM-199. There are many types of these cells, such as cumulus granulosa cells, fibroblasts, trophoblast cells, luteal cells and so on. But there is still a problem of low blastocyst development rate which needs to be further improved.

2 Cloning technique

In biology, cloning refers to a cell or individual that produces a group of cells or individuals with the same genetic material through asexual reproduction. In animal reproduction, it refers to an embryo biotechnology that produces new individuals with identical genetic material without fertilization of sperm and ovum.

2.1 Embryo splitting

Embryo splitting is a technique to manually produce identical twins or polyembryony by microtechnique. It is an important way to expand the source of improved embryos. Its theoretical basis is that each blastomere of morula has the totipotency of developing independently into a new individual.

There are slight differences in the splitting methods for embryos in different developmental stages. Before morula, direct splitting has greater damage to the bigger blastomeres, so separation of blastomeres is often used. For morula or blastocyst, the blastomere is closely connected and the cell boundary gradually disappears, so embryo bisection is mostly used. During blastocyst spltting, inner cell mass should be equally divided. Embryo splitting is successful in many animals, but there are still many problems to futher study. For example, small birth weight, genetic inconsistency, abnormalities and deformities etc.

(二) 核移植

1996年，英国科学家Wilmut等人利用绵羊的乳腺细胞成功克隆了"多莉"，这一划时代的科技成果震动了整个世界，引起了生物学相关领域的一场革命。核移植的基本操作程序见图4-7。

1. 供体核的分离

胚胎克隆的供体核来自早期胚胎，将供体胚胎分散成单个卵裂球。体细胞克隆的供体核为体外传代培养的体细胞，经血清饥饿使细胞处于G_0期。

2.2 Nuclear transplantation

In 1996, British scientist Wilmut successfully cloned Dolly from the mammary gland cells of sheep. This epoch-making scientific and technological achievement shocked the whole world and caused a revolution in biology and related fields. The basic procedures of nuclear transplation are shown in Figure 4-7.

2.2.1 Isolation of donor nucleus

The donor nuclei of embryo cloning come from the early embryos that are dispersed into single blastomeres. The donor nuclei of somatic cell cloning are somatic cells subcultured in vitro. Serum starvation keeps the cells in G_0 phase.

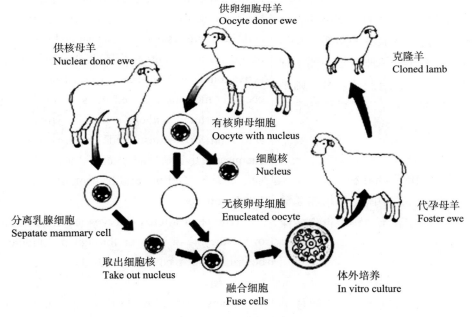

图4-7 核移植程序

Figure 4-7 The procedures of nuclear transplation

2. 卵母细胞的去核

将超数排卵回收的卵母细胞或体外培养成熟的卵母细胞置于含细胞松弛素B和秋水仙胺的培养液中，通过显微操作仪，用吸管吸出第一极体以及处于细

2.2.2 Enucleation of oocyte

The oocytes recovered from superovulation or matured oocytes in vitro are placed in nutrient solution containing cytochalasin B and colchicine. The first polar body, chromosomes at metaphase of cell division

胞分裂中期的染色体和周围的部分细胞质。

3. 细胞融合

将一个卵裂球或体细胞注入去核卵母细胞的卵黄周隙或细胞质中进行融合。细胞融合方法有电融合法和仙台病毒法，后者因融合效果不稳定，且具有感染性，已逐渐被舍弃。电融合具有激活卵母细胞和诱导细胞融合双重作用，即在一定强度的电脉冲作用下，供受体相邻界面的细胞膜发生穿孔，形成细胞间桥从而达到融合，融合率可达96%。

4. 重构胚激活

在正常受精过程中，精子穿过透明带触及卵黄膜时，引起卵子内钙离子浓度升高，卵子恢复正常的细胞周期，启动胚胎发育，这一现象称为激活。在核移植过程中，由于受体卵母细胞处于细胞分裂中期，如果没有激活刺激，则不能恢复细胞周期。人工激活重构胚常利用化学激活法和电激活法。

5. 重构胚的培养与移植

重构胚经体外或中间受体培养至桑葚胚或囊胚，回收胚胎，再进行冷冻保存或胚胎移植。

核移植技术的成功是胚胎工程技术的重大突破，虽然目前克隆动物的成功率较低，但应用前景十分广阔。通过核移植技术，可提高生产效率，加快育种进展，加强濒危动物的保护力度，开展癌细胞相关基因活动的研究。

and some cytoplasm around them are sucked out by a straw under micromanipulator.

2.2.3 Cell fusion

A blastomere or somatic cell is injected into perivitelline space or cytoplasm of the enucleated oocyte for fusion. Cell fusion methods include electrofusion and Sendai virus. The latter has been abandoned gradually because of its unstable fusion effect and infectivity. Electrofusion can activate oocyte and induce cell fusion. Under the electric pulse of certain intensity, the membrane of the donor-recipient interface is perforated and forms intercellular bridges for fusion. The fusion rate is 96%.

2.2.4 Reconstituted embryo activation

During normal fertilization, when sperm passes through the zona pellucida and touches the plasma membrane, the concentration of Ca^{2+} in the ovum increases and the ovum resumes its normal cell cycle, so embryo development is initiated. This phenomenon is called activation. During nuclear transplantation, because of oocyte at metaphase in cell division, the cell cycle can not be restored without activation stimulus. Artificial activation of reconstituted embryos often includes chemical activation and electrical activation.

2.2.5 Culture and transplantation of reconstituted embryos

The reconstituted embryos are cultured to morula or blastocysts in vitro or through the middle receptor before being recovered, and then cryopreserved or transferred.

The success of nuclear transplantation is a great breakthrough in embryo engineering technology. Although the success rate of animal cloning is low at present, its application prospect is very broad. Nuclear transplantation can improve production efficiency, speed up breeding progress, strengthen the protection of endangered animals, and carry out research on gene activity related to cancer cells.

三、性别控制

性别控制是人为干预动物的生殖过程，使雌性动物产出人们期望性别后代的技术。性别控制可充分发挥受性别限制的性状的生产潜力，加快育种进程，防止性连锁疾病的发生。

（一）X、Y 精子的分离

精子分离的主要依据是 X、Y 精子具有不同的物理性质（体积、密度、电荷、运动性）和化学性质（DNA 含量、表面雄性特异性抗原）。

1. 流式细胞仪分离法

流式细胞仪法是目前比较科学、可靠、准确的精子分离方法。其理论依据是：X 精子和 Y 精子的 DNA 含量不同，用荧光染料染色时，DNA 含量高的精子吸收的染料就多，发出的荧光也强，反之发出的荧光就弱。具体方法为：先用 DNA 特异性染料对精子进行活体染色，然后使精子连同少量稀释液逐个通过激光束，探测器可探测精子的发光强度并把不同强弱的光信号传递给计算机。计算机指令液滴充电器使发光强度高的液滴带正电，弱的带负电。然后带电液滴通过高压电场，不同电荷的液滴在电场中被分离，进入两个不同的收集管，正电荷收集管为 X 精子，负电荷收集管为 Y 精子。研究表明，家畜中 X 染色体的 DNA 含量比 Y 染色体高出 3%～4%。

精子分离原理

Principle of sperm separation

3 Sex control

Sex control is a technology that intervenes in animal reproductive process artificially so that females can produce the desired offsprings. Sex control can give full play to the potential of sex-limited character, speed up breeding process and prevent the genetically-linked diseases.

3.1 Separation of X and Y Sperm

The main basis of sperm separation is that X and Y sperm have different physical characteristics (volume, density, electric charge, mobility) and chemical characteristics (DNA content, male-specific antigen on surface).

3.1.1 Flow cytometry separation

At present, flow cytometry is a relatively scientific, reliable and accurate method for sperm separation. The theoretical basis is that the DNA contents of X and Y sperm are different. When using fluorescent dyes, sperm with high DNA content absorbs more dyes and emits stronger fluorescence, conversely, the fluorescence is weak. Specific methods are as follows: first, sperm is dyed using supravital staining method with DNA specific dyes, then sperm passes through the laser beam one by one with a small amount of diluent, the detector can detect the luminous intensity of sperm and transmit different optical signals to the computer. The computer instructs the droplet charger to make the droplets with high luminous intensity charged positively and weak negatively. Then charged droplets are separated into two different collecting tubes by high voltage electric field, X sperm is in positive charge collection tube and Y sperm is in negative charge collection tube. Studies have shown that the DNA content of X chromosome in livestock was 3%-4% higher than that of Y chromosome.

2. 免疫学分离法

免疫学分离法的原理是Y精子质膜携带H-Y抗原,而X精子无此抗原,利用H-Y抗体和H-Y抗原免疫反应检测Y精子,再通过一定的程序分离X精子和Y精子。该法的分离效果不理想,目前很少用。

(二) 控制受精环境

染色体理论并非性别决定机制的全部,外部环境中的某些因素也是性别决定机制的重要条件,例如营养、pH、温度等。由于X、Y两类精子在子宫颈内的游动速度不同,因此到达受精部位与卵子结合的优先顺序不同。另外,Y精子对酸性环境的耐受力比X精子差,当生殖道pH呈弱酸性时,Y精子活力减弱,失去较多与卵子结合的机会,故后代雌性较多。

(三) 胚胎的性别鉴定

哺乳动物早期胚胎的性别鉴定技术现已成熟,鉴定的准确率比较高。目前,早期胚胎性别鉴定最有效的方法主要有核型分析法和分子生物学法。

1. 核型分析法

该法主要操作程序:取少量胚胎细胞,用秋水仙素处理使细胞处于有丝分裂中期,经固定和染色处理,通过显微镜确定其性染色体类型是XX还是XY。此法准确率可达100%,但对胚胎损伤大,且耗时费力,难以在生产中应用。

2. 分子生物学法

此法的理论依据是SRY基因仅存在于Y染色体上,利用分子生物学技

3.1.2 Immunology separation

The principle is that Y sperm plasma membrane carries H-Y antigen, but X sperm does not. Y sperm is detected by immune reaction of H-Y antibody and H-Y antigen, then X sperm and Y sperm are separated by a certain procedure. The effect of this method is not ideal, and is rarely used at present.

3.2 Controlling fertilization environment

Chromosome theory is not the whole mechanism of sex determination. Some factors of external environment are also important for sex determination, such as nutrition, pH and temperature etc. Because of the different swimming speeds of X and Y sperm in the cervix, the priority order of reaching the fertilized site to combine with the ovum is different. In addition, the tolerance of Y sperm to acidic environment is worse than that of X sperm. When the reproductive tract pH is weakly acidic, the vitality of Y sperm decreases and more opportunities to combine with ovum are lost, so the offspring are mostly females.

3.3 Sex identification of embryos

The technology of sex identification of early embryos is fully developed, and the accuracy rate is high. At present, the most effective methods for sex determination of early embryos are karyotype analysis and molecular biology.

3.3.1 Karyotype analysis

Main operating procedures: take a few embryonic cells and treat with colchicine to make them in the metaphase of mitosis. After fixation and dyeing, the type of sex chromosome is determined by microscope as either XX or XY. The accuracy rate can reach 100%, but it causes great damage to embryos and is time-consuming, which makes it difficult to apply in production.

3.3.2 Method of molecular biology

The theoretical basis is that SRY gene only exists on Y chromosome, and the sex of embryo can be deter-

术鉴别胚胎细胞是否存在 *SRY* 基因即可判断出胚胎性别。该法快速、灵敏、简便、准确，对胚胎损伤较小，已广泛应用于胚胎的性别鉴定。

四、胚胎嵌合

胚胎嵌合就是通过显微操作，把两枚或多枚胚胎融合成为一枚复合胚胎，由此而发育成的个体称为嵌合体。胚胎嵌合的方法主要分为卵裂球聚合法和囊胚注入法，前者是将2个以上胚胎的卵裂球相互融合，形成一个胚胎；后者是把一个胚胎的内细胞团注入另一个胚胎的囊胚腔内，使之与原来的内细胞团融合在一起。

目前，已获得鼠、兔、绵羊、山羊、猪、牛等种内嵌合体，以及大鼠-小鼠、绵羊-山羊、牛-水牛和鹌鹑-鸡的种或属间嵌合体。但嵌合体动物的表型性状仅限于一代，不能传递给后代。

五、胚胎干细胞

胚胎干细胞（ESC）是早期胚胎或原始生殖细胞经体外分化抑制培养而获得的可以连续传代的发育全能性细胞系。目前，ESC 已经引起广大学者的关注，对 ESC 的研究也取得了很大进展，相继建立了小鼠（1981年）、仓鼠（1988年）、猪（1990年）、水貂（1992年）、牛（1992年）、兔（1993年）、绵羊（1994年）等的 ESC 系或类 ESC 系。ESC 在功能上具有发育全能性及不断增殖的能力，在生物学领域有着不可估量的应用价值。

mined by using molecular biology technology to identify the existence of SRY gene in embryonic cells. The method is rapid, efficient, simple, accurate and has less damage to embryos. It has been widely used in sex identification of embryos.

4　Embryo chimera

Embryo chimera refers to fusing two or more embryos into a compound embryo by micromanipulation, and the individual developed from it is called chimera. Embryo chimera includes two methods: blastomere aggregation method and blastocyst injection method, the former is to fuse the blastomeres of more than two embryos, the latter is to inject the inner cell mass of one embryo into the blastocele of another embryo to fuse.

At present, intraspecific chimeras (such as rat, rabbit, sheep, goat, pig and cattle) and interspecific chimeras (such as rat-mouse, sheep-goat, cattle-buffalo and quail-chicken) have been obtained. However, the phenotype of chimera is limited to one generation and cannot be transmitted to offsprings.

5　Embryonic stem cells

Embryonic stem cell (ESC) is a continuous totipotent cell line derived from early embryos or primordial germ cells cultured by differentiation inhibition in vitro. At present, ESC has attracted the attention of scholars, and great progress has also been made in the study of ESC. ESC lines or ESC-like lines of mice (1981), hamsters (1988), pigs (1990), mink (1992), cattle (1992), rabbits (1993) and sheep (1994) have been established successively. ESC has the totipotency and ability of continuous proliferation, and it has immeasurable applicational value in the field of biology.

1. ESC 培养体系

这是分离和培养 ESC 的关键环节，目前 ESC 培养体系主要有三种：条件培养体系、分化抑制因子培养体系和饲养层培养体系。

条件培养体系是将细胞培养一段时间后，回收培养液来培养 ESC。分化抑制因子培养体系是将分化抑制因子〔白血病抑制因子（LIF）或白细胞介素-6〕按一定浓度直接添加到细胞培养液中培养 ESC。目前最常用的 ESC 培养体系是饲养层培养体系，饲养层一般由小鼠成纤维细胞经 γ 射线照射或丝裂霉素 C 处理获得，与内细胞团（ICM）或原始生殖细胞（PGC）共同培养即可分离出 ESC。

2. ICM 及 PGC 的选择和分离

许多早期胚胎都可作为建立 ESC 的材料，例如小鼠的桑葚胚、囊胚和扩张囊胚；猪、绵羊、山羊、兔、仓鼠的囊胚；牛的桑葚胚和囊胚。分离 ESC 首先要获得 ICM，然后把 ICM 分散成单个细胞，再放入分化抑制因子培养体系中继续培养。常用的 PGC 分离方法有机械法和消化法，对于分离的 PGC 还要进一步纯化。

3. ESC 的鉴定

当 ICM 增殖后或 PGC 出现胚胎干细胞样克隆后即可传代，2～7 代时进行鉴定。ESC 的鉴定方法主要有形态学鉴定、表面抗原检测、核型分析、体外分化实验等。ESC 必须具有高度的分化潜能，被注入囊胚腔后，能参与内胚层、中胚层和外胚层的形成。体外培养时，分化诱导剂可诱导其定向分化。

5.1 ESC culture system

This is the key to isolate and culture ESC. At present, there are three main ESC culture systems: conditional culture system, differentiation inhibitor culture system and feeder layer culture system.

Conditional culture system refers to the culture of ESC by using recovery culture solution after culturing cells for a period of time. Differentiation inhibitor culture system refers to the culture of ESC by using cell culture solution which is added a certain concentration of leukemia inhibitory factor(LIF) or interleukin-6(IL-6). At present, feeder layer culture system is commonly used. Feeder layer is usually obtained from mouse fibroblasts irradiated by γ-rays or treated with mitomycin C, and then co-cultured with inner cell mass (ICM) or primordial germ cells(PGC), from which ESC can be isolated.

5.2 Selection and separation of ICM and PGC

Many early embryos can be used as materials for the establishment of ESC, such as mouse morula, blastocyst and expanded blastocyst, the blastocyst of pig, sheep, goat, rabbit and hamster, bovine morula and blastocyst. To isolate ESC, ICM should be obtained first, then ICM is dispersed into single cells and cultured in the system of differentiation inhibitors. The common separation methods of PGC are mechanical method and digestion method, the separated PGC needs to be further purified.

5.3 Identification of ESC

When ICM increases or PGC appears embryonic stem cell-like clones, it can be passaged and identified at 2-7 generations. The main identification methods of ESC include morphological identification, surface antigen detection, karyotype analysis and differentiation test in vitro. ESC must have a high differentiation potential and can participate in the formation of endoderm, mesoderm and ectoderm after being injected into blastocoele. When cultured in vitro, differentiation inducers

胚胎干细胞系的建立是胚胎生物技术领域的重大成就，在核移植、嵌合体、转基因动物研究方面进行了广泛的尝试，已经充分体现出胚胎干细胞在加快良种繁育、生产转基因动物、构建哺乳动物发育模型、基因和细胞治疗等方面有广阔的应用前景。

六、转基因技术

转基因技术是用一定方法将目的基因导入受体基因组中或把受体基因组中一段 DNA 切除，从而改变该物种的遗传信息。自 1982 年获得转基因"超级鼠"以来，动物转基因技术已成为当今生命科学中一个发展最快、最热门的领域，相继在兔、羊、猪、牛、鸡等动物获得成功。

转基因技术的主要步骤包括：获取目的基因、构建表达载体、基因导入、鉴定与筛选。其中基因导入是关键环节，基因导入的方法主要包括显微注射法、病毒转染法、精子载体法、ESC 介导法及体细胞核移植法等。

1. 显微注射法

借助显微操作仪，将外源基因直接注入受精卵原核。世界上第一只转基因小鼠就是用这种方法获得的，该技术已广泛应用于制作转基因动物。

can induce its directional differentiation.

Establishment of embryonic stem cell lines is an important achievement in the field of embryonic biotechnology. Extensive attempts have been made in nuclear transplation, chimerism and transgenic animal research. It has broadened application prospects in accelerating breeding of favorable breeds, producing transgenic animals, constructing mammalian development models, gene research and cell therapy etc.

6 Transgenic technology

Transgenic technology refers to changing the genetic information of the species by either introducing the target gene into the receptor genome or removing a segment of DNA from the receptor genome. Since the transgenic super mouse was obtained in 1982, animal transgenic technology has become one of the fastest growing and highly-sought-after fields in life science. It has been successful in rabbits, sheep, pigs, cattle and chickens consecutively.

The main steps of transgenic technology include obtained the target genes, constructing expression vectors, gene introduction, identification and screening. Among them, gene introduction is the key. Its methods include microinjection, virus transfection, sperm vector, ESC mediation and somatic cell nuclear transfer.

6.1 Microinjection

With the help of micromanipulator, exogenous gene is directly injected into the pronucleus of fertilized eggs. The first transgenic mice in the world was obtained in this way. It has been widely used in the production of transgenic animals.

2. 病毒转染法

将目的基因整合到病毒基因组中，然后利用此病毒感染胚胎细胞，即可对胚胎细胞进行遗传转化。该法操作简单、宿主广泛、转染率高，但载体病毒基因有潜在致病性，威胁受体动物的健康安全。

3. 精子载体法

将外源基因片段与获能精子一起孵育，通过受精过程把外源基因导入受精卵（图4-8）。该法最大的优点是方法简单，不需要昂贵复杂的设备，缺点是效果不稳定。

6.2 Virus transfection

Integrate the target gene into the viral genome, and use the virus to infect embryonic cells, then the embryonic cells can be genetically transformed. The method is simple in operation, wide in host and high in transfection rate, but the vector virus gene has potential pathogenicity, which threatens the health and safety of recipients.

6.3 Sperm vector

The exogenous gene fragments are incubated with capacitated sperm, and the exogenous genes are introduced into the fertilized eggs through the fertilization (Figure 4-8). It is simple and does not require expensive and complicated equipments. But it is unstable.

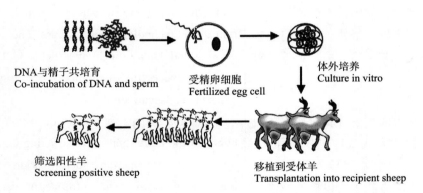

图 4-8　精子载体法

Figure 4-8　Sperm vector

4. ESC 介导法

将外源基因整合到 ESC 基因组中的特定基因位点，通过筛选，把阳性细胞注入受体的囊胚腔内，生产嵌合体动物。当 ESC 分化为生殖干细胞时，外源基因可通过生殖细胞遗传给后代，在第二代获得转基因动物。其理论依据在于囊胚的内细胞团含有尚未分化的胚胎干细胞，将这些干细胞植入正常发育的

6.4 ESC mediation

The exogenous gene is integrated into specific gene locus in ESC genome, and the screened positive cells are injected into the blastocele of the recipient to produce chimaera. When ESC differentiated into germ stem cells, the exogenous gene can be transmitted to offsprings, and the transgenic animals can be obtained in the second generation. Its theoretical basis is that the inner cell mass of blastocyst contains undifferentiated embryonic stem cells, implanting them into the normal

囊胚腔之后，很快便与受体内细胞团聚集在一起，参与正常胚胎的发育。

5. 体细胞核移植法

将外源基因以DNA转染方式导入能进行传代培养的体细胞，再以这些转基因体细胞为核供体，进行动物克隆，获得转基因克隆动物。

转基因工程已广泛应用于农业和医药，甚至与环境保护有密切的关系。在动植物生产方面，可用于进行品种改良和抗病力提高，目前已有百余种转基因植物进入商品化生产，例如水稻、玉米、棉花、大豆等。在医学上，可用于基因治疗和药用蛋白生产。目前市场上的胰岛素、干扰素、白细胞介素-2、生长激素等基因工程药物，均为重组蛋白质或肽类。基因工程药物疗效好、副作用小，已成为各国政府和企业投资研发的热点。将外源正常基因导入靶细胞，以纠正或补偿因基因缺陷和异常引起的疾病，从而达到治疗的目的。

blastocele, they soon gather together with the recipient cells and participate in the development of normal embryos.

6.5 Somatic cell nuclear transfer

The exogenous gene is transfected into subcultured somatic cells by DNA transfection, and then these transgenic somatic cells are used as nuclear donors for animal cloning to obtain transgenic cloned animals.

Transgenic engineering has been widely used in agriculture and medicine industries, and is even closely related to environmental protection. In animal and plant production, it can be used for breed improvement and disease resistance. At present, more than 100 kinds of transgenic plants have entered commercial production, such as rice, corn, cotton, soybean and so on. In medicine, it can be used in gene therapy and pharmaceutical protein production. At present, drugs made by gene engineering such as insulin, interferon, interleukin-2 and growth hormone are all recombinant proteins or peptides. Drugs made by gene engineering have good curative effects and little side-effect, which has greatly attracted governments and enterprises to invest in research and development of this area. In order to treat diseases caused by gene defects and abnormalities, normal exogenous gene is introduced into target cells to correct or compensate for the defected genes.

项目五　妊娠诊断
Project Ⅴ　Pregnancy Diagnosis

项目导学

妊娠是指母畜从受精到分娩的生理过程。母畜配种后应尽早确定其是否妊娠，通过妊娠诊断，加强对已妊娠母畜的饲养管理，使产犊间隔最小化。因此，简便有效的妊娠诊断方法，尤其是早期妊娠诊断方法，受到广大繁殖工作者的关注。

Project Guidance

Pregnancy is the physiological process from fertilization to delivery. Female animals should be determined whether they are pregnant or not as soon as possible after mating. Through pregnancy diagnosis, we should strengthen the feeding management of pregnant to minimize the calving interval. Thus, simple and effective methods of pregnancy diagnosis, especially early pregnancy diagnosis, are favored by breeders.

学习目标

>>> 知识目标

- 理解受精的过程。
- 掌握早期胚胎发育的特点。
- 掌握妊娠识别和胚胎附植的过程。
- 熟悉胎膜和胎盘的结构及特点。
- 了解各种家畜的妊娠期生理。
- 掌握各种母畜的妊娠期。

>>> 技能目标

- 能熟练掌握各种家畜的妊娠诊断方法。
- 根据配种记录，熟练推算各种母畜的预产期。

Learning Objectives

>>> **Knowledge Objectives**

- To understand the process of fertilization.
- To master the characteristics of embryonic development in the early stages.
- To master the process of pregnancy recognition and embryo implantation.
- To know the structure and characteristics of fetal membranes and placentas.
- To understand the physiology of various livestock during pregnancy.
- To master the gestation period for various livestock.

>>> **Skill Objectives**

- To master the diagnosis methods of pregnancy in various livestock.
- According to the mating records, estimate the due dates for various livestock.

相关知识

一、受精生理

受精是指精子和卵子结合形成合子的过程。在这一过程中，精子和卵子经历一系列严格有序的形态和生理变化，单倍体的雌、雄生殖细胞共同构成双倍体的合子。合子是新个体发育的始发点。

（一）配子的运行

在自然情况下，大多数动物的受精发生在母畜输卵管壶腹部。精子由射精部位（或输精部位）、卵母细胞由排出部位到达受精部位的过程，称为配子的运行。了解配子在母畜生殖道内运行及其保持受精能力的时间，对确定适当的配种时间和提高受胎率具有重要的意义。

1. 精子的运行

在自然交配中，雄性动物将精液射入阴道（牛、羊、兔和灵长类动物）或子宫（马、猪、犬和啮齿动物）。射精后，精子在母畜生殖道的运行主要通过子宫颈、子宫和输卵管三个主要部位，最后到达受精部位。

（1）精子在子宫颈内的运行。阴道内射精的动物，如牛、羊，在自然交配时精子存放在阴道内，部分精子借自身运动和黏液向前运动进入子宫，大部分精子进入子宫颈隐窝的黏膜皱襞内，暂时贮存，形成精子库，精子会随子宫颈的收缩活动缓慢释放进入子宫，而死精子可能因纤毛上皮的逆蠕动被推向阴道排出，或被白细胞吞噬而清除。

Relevant Knowledge

1 Fertilization physiology

Fertilization refers to the process in which sperm and ovum combine to form zygotes. During this process, sperm and ovum undergo a series of strictly ordered morphological and physiological changes. The haploid germ cells constitute a diploid zygote. Zygote is the beginning of new ontogeny.

1.1 Transport of gametes

In nature, fertilization of most animals occur in the ampulla of the oviduct of the female. The process of sperm from the ejaculation site (or the insemination site) and oocyte from ovary to the fertilization site is called the transport of gametes. It is important to know the time of gametes running in the reproductive tract of female animals and their ability to maintain fertility, which is imperative for knowing proper mating time and improving conception rate.

1.1.1 Transport of sperm

In natural mating, the male ejaculates semen into the vagina (cattle, sheep, rabbit and primates) or uterus (horse, pig, dog and rodents). After ejaculation, sperm travels through the cervix, uterus and oviduct to reach the fertilization site.

(1) Transport of sperm in the cervix

Animals with endovaginal ejaculation, such as cattle and sheep, store sperm in the vagina when mating naturally. Some sperm enter the uterus by self-movement and mucus forward movement. Most sperm enter the mucosal folds of the cervical recess and are temporarily stored to form a sperm bank. Sperm release slowly into the uterus with the contraction of the cervix. Dead sperm may be pushed to the vagina through the reverse peristalsis of ciliary epithelium and expelled or cleaned up by white blood cells.

子宫颈是精子运行过程中的第一道栅栏，精子经过子宫颈筛选，既保证了运动和受精能力强的精子进入子宫，也防止过多的精子同时涌入子宫。绵羊一次射精将近 30 亿个精子，但能通过子宫颈进入子宫的不足 100 万个。

（2）精子在子宫内的运行。穿过子宫颈进入子宫的精子，在子宫肌收缩作用下，大部分进入子宫内膜腺隐窝中，形成第二个精子库。活力较强的精子从中不断释放，并在子宫肌和输卵管系膜的收缩、子宫液的流动以及精子自身运动等综合作用下，通过宫管连接部进入输卵管。其中一些死精子和活动力差的精子被白细胞吞噬，使精子又一次得到筛选。由于宫管连接部比较狭窄，大量精子滞留于该部，宫管连接部成为精子运行的第二道栅栏。

（3）精子在输卵管中的运行。进入输卵管的精子，在输卵管的收缩、管壁上皮纤毛的摆动作用下，继续前行。在壶峡连接部，精子因峡部括约肌的有力收缩被暂时阻挡，因此，壶峡连接部成为精子到达受精部位的第三道栅栏。各种动物能够到达输卵管壶腹部的精子一般不超过 1 000 个。

精子在母畜生殖道运行中的损耗情况见图 5-1。

精子在母畜生殖道内运行的动力包括射精的力量、子宫颈的吸入作用、母畜生殖道的收缩、生殖道管腔液体的流动以及精子自身的运动等。精子由射精部位到达受精部位所需的时间与母畜的生理状况有关，一般为 20min 左右。

The cervix is the first barrier in the process of sperm transport. The cervix ensures motile and fertilized sperm enter the uterus, and also prevents excessive sperm from entering the uterus at the same time. For example a ram ejaculates nearly 3 billion sperm at a time, but less than 1 million can enter the uterus through the cervix.

（2）Transport of sperm in uterus

The sperms passing through the cervix enter the uterus. Under the contraction of uterine muscles, most of sperms enter the endometrial glands, forming a second sperm bank. Vigorous sperm is released continuously from the endometrial glands, and enters the oviduct through the uterotubal junction under the influence of the contraction of uterine muscle and mesosalpinx, flow of uterine fluid and sperm motility. The dead and poor motility sperms are swallowed by white blood cells, which further selects the viable sperm. Owing to the narrowness of the uterotubal junction, a large number of sperms remain in the uterotubal junction, which becomes the second barrier for sperm transport.

（3）Transport of sperm in the oviduct

The sperms in the oviduct move forward continuously by the contraction of the oviduct and swing of tube wall epithelial cilia. They are temporarily blocked due to the strong contraction of the isthmic sphincter in the junction of ampulla and isthmus, which becomes the third barrier for sperms to reach the fertilization site. Generally, no more than 1 000 sperms can reach the ampulla of oviduct.

The loss of sperm in reproductive tracts of female animals is shown in Figure 5-1.

The motive force of sperm in the reproductive tract of female animals includes the force of ejaculation, the absorption of cervix, the contraction of reproductive tract, the flow of liquid in the reproductive tract and the movement of sperm. The time required for sperms the fertilization site from the ejaculation site is related to the physiological conditions

精子在母畜生殖道的存活时间一般为1～2d，牛为15～56h，猪约为50h，羊约为48h，马约为6d。精子维持受精能力的时间短于存活时间，牛约为28h、猪约为24h、绵羊为30～36h，马为5～6d。

of the female animals, generally about 20 minutes. The survival time of sperm in the reproductive tract of the female animals is generally 1-2 days, the cow is 15-56 hours, the pig is about 50 hours, the sheep is about 48 hours, and the horse is about 6 days. The time of sperm fertilization window is even shorter than the survival time, which is about 28 hours for cattle, 24 hours for pigs, 30-36 hours for sheep, and 5-6 days for horses.

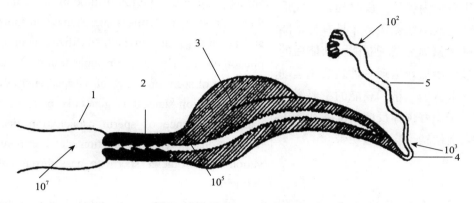

图 5-1　精子运行中的损耗
Figure 5-1　Sperm loss in transport
1. 阴道　2. 子宫颈　3. 子宫　4. 宫管连接部　5. 输卵管
1. vagina　2. cervix　3. uterus　4. uterotubal junction　5. oviduct

2. 卵子的运行

接近排卵时，输卵管伞充分开放、充血，紧贴于卵巢表面。输卵管伞黏膜上摆动的纤毛将卵巢排出的卵子接纳入喇叭口。猪和马的伞部发达，卵子易被接受；牛、羊因伞部不能完全包围卵巢，有时造成排出的卵子落入腹腔，再靠纤毛摆动形成的液流将卵子吸入输卵管。被伞部接纳的卵子，借输卵管管壁纤毛的颤动、平滑肌的收缩以及腔内液体的作用，被运送到受精部位。卵子在输卵管内运行的时间，牛约为80h，猪约为50h，绵羊约为72h。但卵子保持受精能力的时间，一般为1d以内，牛为18～20h，猪为8～12h，绵羊为12～16h。

1.1.2　Transport of oocyte

Before ovulation, the oviduct umbrella is fully open and congested, and close to the ovarian surface. The waving cilia on the mucosa of the oviduct umbrella receives the oocyte from the ovary. Owing to the developed oviduct umbrellas, the oocyte of pigs and horses are easily accepted. Because of the oviduct umbrellas of cattle and sheep can not completely encircle the ovaries, sometimes the oocyte falls into the abdominal cavity, but the fluid flowing through the cilia can suck the oocyte back into the oviduct. The time of oocyte running in the oviduct is about 80 hours for cattle, 50 hours for pigs and 72 hours for sheep. However, the time window for fertilization is generally less than one day, which is 18-20 hours for cattle, 8-12 hours for pigs, and 12-16 hours for sheep.

（二）配子在受精前的准备

受精前，哺乳动物的精子和卵子都要经历一个进一步生理成熟的阶段，才能顺利完成受精过程，并为受精卵的发育奠定基础。

1. 精子的获能

刚排出的精子没有受精能力，只有在雌性生殖道内经历一段时间，完成形态和生理上的一些变化，才具有受精能力，这种现象称为精子获能。获能后的精子耗氧量增加，呈现一种超活化运动状态。一般认为，精子获能的主要意义在于使精子做好顶体反应的准备和精子超活化，促使精子穿越透明带。

对于子宫射精型的动物，精子获能开始于子宫，结束于输卵管；对于阴道射精型的动物，精子获能开始于阴道，当子宫颈开放时，流入阴道的子宫液可使精子获能，但获能最有效的部位是子宫和输卵管。精子获能也可在体外人工培养液中完成。精子在雌性生殖道内获能的时间，因动物种类不同而异，一般牛为3～4h、猪为3～6h、绵羊为1.5h、兔为5～6h。

精子获能是一个可逆过程。获能的精子，若重新置于精清中便又失去受精能力，这种现象称为去能。如果去能的精子再回到母畜生殖道内，可再次获能。在哺乳动物，去能因子无种间特异性。

获能后的精子，在受精部位与卵子相遇，会出现顶体帽膨大，精子质

1.2 Preparation of gametes before fertilization

Before fertilization, the sperm and oocyte of mammalians have to undergo a stage of further physiological maturation to successfully complete the fertilization process and lay a good foundation for the development of zygote.

1.2.1 Sperm capacitation

The freshly ejaculated sperm has no fertilization ability, only after a period of time in the female reproductive tract, completing some changes in morphology and physiology, it can have fertilization ability. This phenomenon is called sperm capacitation. Oxygen consumption of sperm increased after capacitation, showing a hyperactivation state. It is generally believed that the main significance of sperm capacitation lies in the preparation of acrosome reaction and sperm superactivation, and promotes sperm to pass through the zona pellucida.

In uterine ejaculation animals, sperm capacitation begins in the uterus and ends in the oviduct. In vaginal ejaculation animals, sperm capacitation begins in the vagina. When the cervix is open, the uterine fluid flowing into the vagina can make sperm capacitated, but the most effective sites of sperm capacitation are the uterus and oviduct. Sperm capacitation may also be accomplished in an artificial medium in vitro. The time of sperm capacitation in the female reproductive tract varies among species. For the cattle it is 3-4 hours, the pig is 3-6 hours, the sheep is 1.5 hours, and the rabbit is 5-6 hours.

Sperm capacitation is a reversible process. Capacitive sperm will lose fertility if they are replaced in the seminal fluid. This phenomenon is called decapacitation. If the decapacitive sperm returns to the reproductive tract of the female animal, it can be capacitated again. In mammals, decapacitation factors have no interspecific specificity.

When the capacitated sperm meets the oocyte at the fertilization site, acrosome cap enlarges, spermatozoal

膜和顶体外膜融合而形成许多泡状结构，最后由顶体内膜和顶体基质释放出顶体酶系，这一过程称为顶体反应。顶体反应为精子穿越卵子并与之融合奠定了基础。

2. 卵子的成熟

刚排出的卵子，在进入输卵管壶腹部前尚不具备受精能力，如猪和羊排出的卵子为刚完成第一次减数分裂的次级卵母细胞，马和犬排出的卵子仅为初级卵母细胞。卵子需在输卵管内进一步成熟，达到第二次减数分裂中期，才具备被精子穿透的能力。同时，卵子的皮质颗粒不断增加，并向卵的周围移动，此时透明带出现精子受体，卵黄膜发生亚显微结构变化。

（三）受精过程

受精过程指精子和卵子相结合的生理过程。哺乳动物的受精过程主要包括以下 5 个阶段（图 5-2）。

plasma membrane and outer acrosome membrane fuse to form many vesculation. At last, acrosome enzymes are released from inner acrosome membrane and acrosome matrix. This process is called acrosome reaction. This reaction lays the foundation for sperm to cross and fuse with the oocyte.

1.2.2 Oocyte maturation

The newly expelled oocytes do not have the ability to fertilize before entering the ampulla of the oviduct, for example, the oocytes from pigs and sheep are secondary oocytes that have just completed their first meiotic division, and the oocytes from horses and dogs are only primary oocytes. Oocytes need to mature in the oviduct to reach second meiotic metaphase before they can be penetrated by sperm. At the same time, the cortical granules of the oocytes increase continuously and move around the oocytes. At this time, sperm receptor in zona pellucida appears, and microstructural changes occur in yolk membrane.

1.3 Fertilization process

Fertilization refers to the physiological process of sperm fusing with oocyte. The fertilization process of mammals mainly includes the following five stages (Figure 5-2).

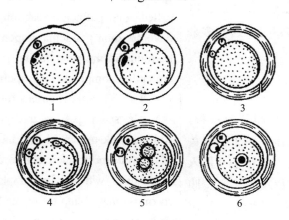

图 5-2 受精过程

Figure 5-2 Fertilization process

1. 精子与透明带接触　2. 精子进入透明带　3. 精子进入卵黄　4、5. 雄、雌原核形成　6. 配子配合

1. sperm contacts with zona pellucida　2. sperm enters zona pellucida　3. sperm enters yolk

4，5. prokaryotic formation of male and female　6. cyngamy

1. 精子穿过放射冠

放射冠是包围在卵子透明带外面的卵丘细胞群。受精前大量精子包围着卵细胞，经顶体反应的精子释放出透明质酸酶，溶解放射冠的胶样基质，使精子接触到透明带。此过程需要大量精子的共同作用，此时卵子对精子无选择性，即不存在种间特异性。

2. 精子穿过透明带

接触到透明带的精子，很快与透明带上的精子受体结合。精子受体具有明显的种属特异性，只有同种动物的精子才可与其受体结合。精子释放顶体素（酶）将透明带溶出一条通道，精子借自身运动穿过透明带。

当精子穿过透明带触及卵黄膜时，将处于休眠状态的卵子"激活"，卵黄膜发生收缩，由卵黄释放出某种物质，传播到卵的表面及卵黄周隙，阻止后来的精子再进入透明带，这一变化称为透明带反应。兔的卵子无透明带反应，可在卵黄周隙内发现许多精子，这些多余的精子称为补充精子。

3. 精子进入卵黄

穿过透明带的精子与卵黄膜接触，卵黄膜表面的微绒毛包围住精子，卵黄膜随之与精子质膜融合，将精子"拖入卵内"。当精子进入卵黄膜时，卵黄膜立即发生卵黄紧缩、卵黄膜增厚、排出部分液体进入卵黄周隙等变化，拒绝其他精子再进入卵黄，这种现象称为卵黄膜封闭作用或多精子入卵阻滞。

1.3.1 Sperm penetrates the corona radiata

The corona radiata is a cumulus cell group surrounding the zona pellucida. Before fertilization, a large number of sperm surround the oocyte. After the acrosome reaction, the sperm releases hyaluronidase, dissolves the colloidal matrix of the corona radiata, and makes the sperm touch the zona pellucida. This process requires the interaction of a large number of sperms, at this moment the oocyte has no sperm selectivity, that is, there is no interspecific specificity.

1.3.2 Sperm passes through zona pellucida

Sperm exposed to zona pellucida quickly binds to sperm receptors in zona pellucida. Sperm receptors have distinct species specificity and only the sperm of the same species can bind to their receptors. The sperm dissolves a channel through zona pellucida by release of acrosin(enzyme), and the sperm moves through the zona pellucida by itself.

When sperm passes through the zona pellucida and touches the yolk membrane, the dormant oocyte is activated and the yolk membrane contracts, releasing a substance from the yolk, which spreads to the surface and perivitelline space, and prevents the subsequent sperm from entering the zona pellucida. This change is called zona pellucida reaction. The oocytes of rabbits have no zona pellucida response, so many sperms can be found in the perivitelline space. These extra sperms are called complementary sperm.

1.3.3 Sperm enters the yolk

The sperm passing through zona pellucida contacts with the yolk membrane, and the microvilli on the surface of the yolk membrane embrace the sperm. The yolk membrane fuses with the sperm plasma membrane, and the sperm is dragged into the ovum. When the sperm enters the yolk membrane, the yolk membrane immediately undergoes changes, such as yolk contraction, yolk membrane getting thicker, and part of the fluid being discharged into the perivitelline space, and other spems can't enter the yolk. This phenomenon is called vitelline block.

项目五 妊娠诊断
Project V Pregnancy Diagnosis

4. 原核形成

精子进入卵黄后，头部膨大，尾部脱落，出现核膜和核仁，形成雄原核。精子进入卵黄后不久，卵子排出第二极体，完成第二次成熟分裂，逐渐出现核膜和核仁，形成雌原核。原核形成后，不断发育、变大。猪的两个原核大小相似，其他家畜雄原核略大于雌原核。

5. 配子配合

雄原核和雌原核经充分发育，相向移动，彼此接触，融合在一起，核仁、核膜消失，两组染色体合并成一组，这一过程称为配子配合。配子配合完成后，受精至此结束。

二、妊娠生理

妊娠是指母畜从受精开始，经过胚胎的生长发育，直至胎儿成熟产出体外的生理变化过程。

（一）胚胎的早期发育

从受精卵第一次卵裂至发育成原肠胚的过程，称为胚胎的早期发育。根据形态特征，可将早期胚胎的发育分为桑葚胚、囊胚和原肠胚三个阶段（图5-3）。

1. 桑葚胚

受精卵形成后立即在透明带内进行卵裂。第一次卵裂，合子一分为二，形成两个卵裂球。之后，胚胎继续卵裂，但每个卵裂球并不一定同时进行分裂，故可能出现单数个细胞的时期。当卵裂球达到16～32个细胞时，由于透明带

1.3.4 Pronucleus formation

When the sperm enters the yolk, the head enlarges, the tail falls off, and the nuclear membrane and nucleolus appear, forming the male pronucleus. Shortly after the sperm enters the yolk, the oocyte discharges the second polar body and completes the second mature division, and the nuclear membrane and nucleolus appears gradually, forming the female pronucleus. After the formation of pronucleus, they grow and become bigger. The male pronucleus are always slightly larger than the female pronucleus, but the two pronucleus of pigs are similar in size.

1.3.5 Cyngamy

The male and female pronucleus are fully developed, moving towards each other and fusing together, then the nucleolar and the membrane disappear. The two groups of chromosomes merge into a group. This process is called cyngamy. After cyngamy is completed, the fertilization ends.

2 Pregnancy physiology

Pregnancy refers to the physiological changes of the female animals from fertilization to the growth and development of the embryo, until the fetus matures enough to be delivered.

2.1 Early embryonic development

The process of the development from the first cleavage of a oosperm to the gastrula is called the early embryonic development. According to morphological characteristics, the early embryonic development can be divided into three stages: morula, blastocyst and gastrula(Figure 5-3).

2.1.1 Morula

The cleavage occurs immediately in the zona pellucida after the formation of the oosperm. In the first cleavage, the zygote is divided into two parts, forming two blastomeres. After that, the cleave of the embryo continues, but each blastomere does not divide at the same time, so there may be an odd number of cells ap-

的限制，卵裂球在透明带内形成致密的细胞团，形似桑葚，称为桑葚胚。这一时期的变化主要在输卵管内完成，依赖自身卵黄获取营养。

早期胚胎的每个卵裂球都具有发育成为一个新个体的全能性，利用此特性可进行胚胎分割和移植。

peared. When the blastomeres reaches 16-32 cells, due to the restriction of zona pellucida, it forms a dense cell mass which shapes like a mulberry in the zona pellucida, hence its name morula. This period is mainly completed in the oviduct, depending on their own yolk for nutrition.

Each blastomere of an early embryo has the totipotency of developing into a new individual, which can be used for embryo splitting and transplantation.

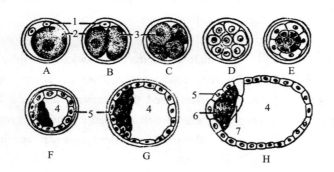

胚胎早期发育
Early development of embryonic

图 5-3　受精卵的发育
Figure 5-3　Development of oosperm
A. 合子　B. 2 细胞期　C. 4 细胞期　D. 8 细胞期　E. 桑葚期　F～H. 囊胚期
1. 极体　2. 透明带　3. 卵裂球　4. 囊胚腔　5. 滋养层　6. 内细胞团　7. 内胚层
A. zygote　B. 2 cells　C. 4 cells　D. 8 cells　E. morula　F-H. blastocyst
1. polar body　2. zona pellucida　3. blastomere　4. blastocoel
5. trophoblast　6. inner cell mass　7. endoderm

2. 囊胚

桑葚胚继续发育，逐渐在细胞团中出现充满液体的小腔，称为囊胚腔，此时的胚胎称为囊胚。随着细胞的分裂，囊胚腔不断扩大，最终一些细胞被挤在腔的一端，细胞密集成团，称为内细胞团；而另一些细胞沿着透明带的内壁排列扩展，构成囊胚腔的壁，这一单层细胞称为滋养层。在囊胚阶段，细胞开始分化，内细胞团将来发育为胎儿，滋养层将发育为胎膜和胎盘。囊胚后期，透

2.1.2 Blastocyst

The morula continues to develop, and a small cavity filled with liquid gradually appears in the cell mass, which is called a blastocyst cavity. The embryo at this time is called a blastocyst. With cell division, blastocyst cavity expands continuously, and finally some cells are squeezed into one end of the cavity, which is called inner cell mass, while others are arranged along the inner wall of zona pellucida to form the wall of blastocyst cavity, which is called trophoblast. During the blastocyst stage, the cells begin to differentiate, the inner cell mass will develop into the fetus, and the trophoblast

明带崩解,囊胚体积迅速增大,变为扩张囊胚。囊胚阶段主要从子宫乳获取营养物质。

3. 原肠胚

囊胚进一步发育,出现了内、外两个胚层,此时的胚胎称为原肠胚。原肠胚继续发育,在滋养层(即外胚层)和内胚层之间出现了中胚层,中胚层又分化为体壁中胚层和脏壁中胚层,两个中胚层之间的腔隙,构成以后的体腔。三个胚层的建立和形成,为胎膜和胎体各类器官的分化奠定了基础。

(二)妊娠识别

在妊娠早期,胚胎即能产生某种化学因子(激素)作为妊娠信号传给母体,母体随即产生相应的生理反应,以识别或确认胚胎的存在。由此胚胎和母体之间建立起密切的联系,这一过程称为妊娠识别。妊娠识别的实质是胚胎产生抗溶黄体物质,作用于母体的子宫或黄体,阻止或抵消 $PGF_{2\alpha}$ 的溶黄体作用,使黄体变为妊娠黄体,维持母畜妊娠。

妊娠识别后,母畜即进入妊娠的生理状态,但各种动物妊娠识别的时间不同,猪为受精后 10~12 d、牛为 16~17 d、绵羊为 12~13 d、马为 14~16 d。

(三)胚胎的附植

早期胚胎在子宫内游离一段时间后,体积越来越大,其活动逐渐受到限制,位置逐渐固定下来,胚胎的滋养层和子宫内层膜逐渐建立起组织和生理上

will develop into the fetal membrane and placenta. At the later stage of blastocyst, the zona pellucida disintegrates, and the blastocyst increases rapidly and becomes expanded blastocyst. At this stage, blastocysts obtain nutrients mainly from the uterine milk.

2.1.3 Gastrula

With the development of blastocyst, endoderm and ectoderm appear. At this time, the embryo is called gastrula. The gastrula continues to develop, mesoderm appears between trophoblast (ectoderm) and endoderm. The mesoderm differentiates into somatic mesoderm and splanchnic mesoderm. The gap between the two mesoderms forms the coelom. The formation of three parts lays a foundation for the differentiation of various organs.

2.2 Pregnancy recognition

In the early stage of pregnancy, the embryo can produce some chemical factors (hormones) as a pregnancy signal to the mother, then maternal physiological reactions occur to identify or confirm the existence of embryo. Thus, a close relationship is established between the embryo and the mother. This process is called pregnancy recognition. The essence of pregnancy recognition is that the embryo produces antiluteolytic substance, which acts on the uterus or corpus luteum and prevents or counteracts the luteolytic effect of $PGF_{2\alpha}$, transforming the corpus luteum into the corpus luteum of pregnacy, and maintains the maternal pregnancy.

After pregnancy recognition, the female enters the physiological state of pregnancy. However, the time of pregnancy recognition is different in various animals. It is 10-12 days after fertilization for the pigs, 6-17 days for cattle, 12-13 days for sheep, and 14-16 days for horses.

2.3 Embryo implantation

After a period of intrauterine dissociation, the early embryo becomes larger and larger, its activity is limited and the position is gradually fixed. The tissual and physiological links are established gradually between the trophoblast and endometrium. This process

的联系，这一过程称为附植。

1. 附植时间

胚胎附植是一个逐渐发生的过程，各种家畜胚胎的准确附植时间差异较大。胚胎附植的时间大体为：牛为受精后40～45 d，马为90～105 d，猪为25～26 d，绵羊为28～35 d。

2. 附植部位

胚胎在子宫内附植时，通常都是寻找对胚胎发育最有利的位置。所谓有利，一是指子宫血管稠密的地方，可以提供丰富的营养；二是距离均等，避免拥挤。胚胎在子宫中附植的位置，因动物种类不同而异。牛、羊怀单胎时，胚胎多在排卵侧子宫角下1/3处附植，双胎时则均匀分布于两侧子宫角；马怀单胎时，胚胎常迁至对侧子宫角基部附植；猪的多个胚胎平均等距离分配在两侧子宫角。

（四）胎膜和胎盘

1. 胎膜

胎膜是胎儿的附属膜，是卵黄囊、羊膜、绒毛膜、尿膜和脐带的总称（图5-4）。其作用是与母体子宫黏膜交换养分、气体及代谢产物，对胎儿的发育极为重要。在胎儿出生后，胎膜即被摒弃，所以是一个暂时性器官。

（1）卵黄囊。哺乳动物胚胎发育初期都有卵黄囊的发育，其上分布有稠密的血管，是早期胚胎主要的营养器官，起着原始胎盘的作用。随着胎盘的形成，卵黄囊逐渐萎缩，最后只在脐带中留下一点遗迹。

is called implantation.

2.3.1 Implantation time

Embryo implantation is a gradual process, and the accurate implantation time of various livestock embryos has enormous differences. The time of embryo implantation is roughly 40-45 days after fertilization for cattle, 90-105 days for horses, 25-26 days for pigs, and 28-35 days for sheep.

2.3.2 Implantation site

When the embroy implanting in uterus, it usually looks for the most favorable position for embryonic development. The favorable position which can provide abundant nutrition is where blood vessels are dense in uterine and with equal distance from each other to avoid crowding. The implantation site of embryos in the uterus varies according to animal species. When cows and sheep have a single fetus, the embryo is often attached to 1/3 down the uterine horn on the ovulation side, twins are evenly distributed on both sides of the uterine horn. When the horses have a single fetus, the embryo is often transferred to the base of the contralateral uterine horn. When the pigs have multiples, the embryos are evenly distributed on both sides of the uterine horn.

2.4 Embryonic membrane and placenta

2.4.1 Embryonic membrane

The embryonic membrane is the accessory membrane of the fetus. It is formed of the yolk sac, amnion, chorion, allantois and umbilical cord (Figure 5-4). Its role is to exchange nutrients, gases and metabolites with maternal uterine mucosa, which is very important for fetal development. After the birth of the fetus, it is discarded, so it is a temporary organ.

(1) Yolk sac

The development of yolk sac also starts at the early age of embryo development, it is covered with dense blood vessels. It is the main nutritional organ of early embryos and plays the role of primitive placenta. With the formation of the placenta, the yolk sac atrophies gradually, leaving only a trace in the umbilical cord.

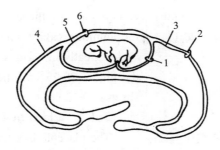

图 5-4 猪的胎膜
Figure 5-4 Embryonic membrane of pigs
1. 尿膜羊膜 2. 尿膜绒毛膜 3. 尿膜 4. 绒毛膜 5. 羊膜 6. 羊膜绒毛膜
1. urinary amnion 2. urinary chorion 3. allantois 4. chorion 5. amnion 6. amniochorion

（2）羊膜。羊膜是包裹在胎儿外面的一层透明薄膜，在胎儿脐孔处和胎儿皮肤相连。羊膜闭合为羊膜腔，其内含有羊水，胎儿即漂浮在羊水中，对胎儿起缓冲作用。

（3）尿膜。尿膜闭合为尿囊，尿囊通过脐带中脐尿管与胎儿膀胱相连，内含尿水，相当于胚体外的临时膀胱。尿膜上分布有大量来自脐动脉、脐静脉的血管。猪、牛、羊的尿囊在胎儿腹侧和两侧半包围着羊膜囊，马的尿囊完全包裹着羊膜囊。

（4）绒毛膜。绒毛膜是胚胎的最外一层膜，表面覆盖绒毛，嵌入子宫黏膜腺窝，形成胎儿胎盘的基础。除马、驴、兔外，其他家畜的绒毛膜均有部分与羊膜接触，形成羊膜绒毛膜。

（5）脐带。脐带是胎儿和胎盘联系的纽带，被覆羊膜和尿膜。脐带随胚胎的发育逐渐变长，使胚体可在羊膜腔中自由移动。脐带的长度种间差异较大，

(2) Amnion

Amnion is a transparent membrane wrapped outside the fetus, which connects with the fetal skin at the umbilical perforation. The amnion is closed as the amniotic cavity, which contains amniotic fluid. The fetus floats in the amniotic fluid which acting as a buffer.

(3) Allantois

Allantois is closed as an allantoic sac, which connects with the fetal bladder through the urachus and contains urine, is equivalent to the temporary bladder outside the embryo. There are a large number of vessels from umbilical artery and umbilical vein on the allantois. The allantoic sac of pigs, cattle and sheep are surrounded by amniotic sac on the ventral side and sides of the fetus, and the allantoic sac of horses are wrapped entirely in the amniotic sac.

(4) Chorion

Chorion is the outermost layer of the embryo, which is covered with villi and is embedded in the glandular fossa of the uterine mucosa to form the basis of the fetal placenta. Apart from horses, donkeys and rabbits, the chorion of other livestock contacts with the amnion to form the amniochorion.

(5) Umbilical cord

Umbilical cord is the link between fetus and placenta, covered with amnion and allantois. The umbilical cord gradually lengthens with the development of the

猪平均为 20～25 cm，牛为 30～40 cm，羊为 7～12 cm，分娩时多数自行断裂。

2. 胎盘

胎盘是由尿膜绒毛膜和妊娠子宫黏膜结合在一起的组织，其中尿膜绒毛膜部分称为胎儿胎盘，而子宫黏膜部分称为母体胎盘。

（1）胎盘的类型。根据绒毛膜表面绒毛的分布将胎盘分为弥散型、子叶型、带状和盘状 4 种类型（图 5-5）。

embryo, so that the embryo body can move freely in the amniotic cavity. The length of the umbilical cord varies greatly in various species. The average length of the pig is 20-25 cm, the cattle is 30-40 cm, and the sheep is 7-12 cm.

2.4.2 Placenta

The placenta is a tissue that combines the urinary chorion with the pregnant uterine mucosa. The part of the urinary chorion is called the fetal placenta, and the part of the uterine mucosa is called the maternal placenta.

（1）Types of placenta

According to the distribution of villi on the chorionic surface, the placenta can be divided into four types: diffuse placenta, cotyledonary placenta, zonary placenta, discoid placenta (Figure 5-5).

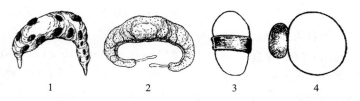

图 5-5 胎盘的类型

Figure 5-5　Types of placenta

1. 弥散型胎盘　2. 子叶型胎盘　3. 带状胎盘　4. 盘状胎盘

1. diffuse placenta　2. cotyledonary placenta　3. zonary placenta　4. discoid placenta

①弥散型胎盘。胎盘绒毛膜上的绒毛分散而均匀地分布在整个绒毛膜表面，如马、驴、猪等。子宫黏膜上皮形成陷窝，绒毛深入到陷窝内。此类胎盘构造简单，结合不牢，易发生流产；分娩时出血较少，胎衣易于脱落。

②子叶型胎盘。以反刍动物牛、羊为代表。尿膜绒毛膜上的绒毛呈丛状分布，相应嵌入母体子宫黏膜上皮的陷窝中。此类胎盘结合紧密，产后易出现胎衣不下。

①Diffuse placenta

The villi of placenta are distributed evenly on the surface of the entire chorionic surface, such as horses, donkeys and pigs. The epithelium of the uterine mucosa forms a lacuna, and the villi penetrate into the lacuna. This type of placenta is simple in structure, weak in combination, and prone to miscarriage, less bleeding during delivery, and easy to fall off the placenta.

②Cotyledonary placenta

Cotyledonary placenta is typically seen in ruminant animals such as cattle and sheep. The villus on the urinary chorion are clustered, and correspondingly embedded in the lacuna of the epithelium of maternal uterine mucosa. This

③带状胎盘。绒毛集中于绒毛膜的中央，呈环带状，故称带状胎盘。犬、猫等食肉动物为此类胎盘。

④盘状胎盘。在胎儿发育过程中，绒毛集中于一个圆形的区域，呈圆盘状。人和灵长类属此类型。

（2）胎盘的功能。胎盘是一种功能复杂的器官，是母体和胎儿相连接的纽带，具有物质运输、合成及分解代谢、分泌激素及免疫等多种功能。

（五）妊娠生理

母畜妊娠期间，由于卵巢上妊娠黄体的存在和孕体的发育，内分泌系统出现明显的变化，使母畜的生殖器官和整个机体都出现了特殊变化，这些变化对妊娠诊断有很好的参考价值。

1. 全身的变化

母畜妊娠后，食欲增加，体重增加，被毛光润，性情温顺。妊娠中后期，由于胎儿增长迅速，尽管母畜食欲增强，但仍入不抵出，膘情有所下降。妊娠末期，胎儿对钙、磷等矿物质需要量增多，若不及时补充，母畜容易出现后肢跛行，牙齿磨损较快。妊娠中后期（马、牛5个月、羊3个月、猪2个月以后），孕畜腹部膨大，排粪（尿）次数增多，行动谨慎，容易疲劳和出汗。

kind of placenta is tightly combined, and the placenta is prone to retain after delivery.

③Zonary placenta

The villi concentrate in the center of the chorion and in a shape of ring-like, so this kind of placenta is called zonary placenta, typical examples including those of dogs, cats and other predators.

④Discoid placenta

During the development of the fetus, the villi of the placenta concentrate in a circular area, which is discoid, such as those of human beings and primates.

（2）Function of placenta

The placenta is a organ with complex function and serves as the link between the mother and fetus. It has many functions, such as material transportation, synthesis, catabolism, hormone secretion and immunity.

2.5 Physiology of pregnancy

During pregnancy, the endocrine system changes obviously due to the existence of corpus luteum and the development of conceptus. Special changes have taken place in the reproductive organs and the whole body of the female animals. These changes provide a good reference value for the pregnancy diagnosis.

2.5.1 Changes of the body

After pregnancy, appetite and weight increase, fur coat is smooth and the temperament of the dam becomes gentle. In mid-late pregnancy, due to the rapid growth of the fetus, although the dam eat more, but that is not enough for the nutrient supply, so that the fatness often declines. At the end of pregnancy, the fetus needs more minerals such as calcium and phosphorus. If not supplemented in time, health problems such as limping hind limps, fast teeth wearing could appear. In mid-late pregnancy (5 months for horse and cattle, 3 months for sheep, 2 months for pigs), maternal abdomen distends, the frequency of excretion and urine increases, the movement is cautious, and easy to fatigue and sweating.

2. 生殖器官的变化

（1）卵巢。母畜妊娠后，卵巢上的妊娠黄体质地较硬，比周期黄体略大，持续存在于整个妊娠期，分泌较多孕酮，以维持妊娠。妊娠早期，卵巢偶有卵泡发育，致使孕后发情，但多数不排卵而退化、闭锁。随着胎儿体积增大，卵巢随子宫沉入腹腔。

（2）子宫。妊娠期间，子宫通过增生、生长和扩展的方式以适应胎儿生长的需要。妊娠前半期，子宫体积的增长主要是子宫肌纤维增生、肥大，后半期由于胎儿生长和胎水增多，子宫壁扩张变薄。由于子宫重量增加，并向前下方垂入，因此至妊娠的后半期，一部分子宫被拉入腹腔，但至妊娠末期，由于胎儿增大，又会被推回至骨盆腔前缘。

子宫颈在妊娠期间收缩紧闭。子宫颈内腺体数目增加，分泌的黏液浓稠，充塞在颈管内形成栓塞，称为子宫栓。子宫栓可防止外界的异物和微生物进入子宫，有保胎作用。牛的子宫颈分泌物较多，妊娠期间有子宫栓更新现象，马、驴的子宫栓较少。子宫栓在分娩前液化排出。

（3）子宫动脉。妊娠期内子宫血管变粗，动脉内膜增厚，且与动脉的肌层联系变疏松，血液流动时出现的脉搏由原来清楚的跳动变为间隔不明显的流水

2.5.2 Changes of reproductive organs

(1) Ovary

After pregnancy, the corpus luteum of pregnancy on the ovary is hard, slightly larger than the corpus luteum of the cycle, which persists throughout the pregnancy period and secretes more progesterone to maintain pregnancy. In the early pregnancy, follicles are occasionally developed in ovaries, resulting in estrus after pregnancy, however, most of them do not ovulate but degenerating into atresia. With the increase of fetal size, ovaries sink into abdominal cavity with uterus.

(2) Uterus

During pregnancy, the uterus adapts to the fetal growth by proliferating, growing and expanding. During the first half of pregnancy, the increase of uterus is mainly due to the hypertrophy of uterine myofibrils. In the second half of pregnancy, uterine wall will dilate and thin due to the fetal growth and increased fetal fluid. As the weight of the uterus increases and falls forward and downward, some of the uterus is pulled into the abdominal cavity in the second half of pregnancy, but at the end of pregnancy, due to the fetal enlargement, it is pushed back to the front of the pelvic cavity.

The cervix is constricted during pregnancy. The number of glands in the cervix increases, and the secretion of mucus is thick, which fills the cervical canal and causes embolism, known as uterine embolism. Uterine embolism can prevent miscarriage by preventing foreign matter and microorganisms from entering the uterus. There are more cervical secretions in cattle, with uterine embolism renewal during pregnancy, but there are fewer uterine embolism in horses and donkeys. Uterine embolism is liquefied and discharged before delivery.

(3) Uterine artery

During pregnancy, uterine blood vessels and endarterium became thicker. The connection between endarterium and arterial muscular layer becomes looser. The pulsation of blood flow changes from the clear to the

样颤动，称为妊娠脉搏（孕脉）。这是妊娠的特征之一，不同母畜孕脉的强弱及出现时间不同。至妊娠末期，牛、马的子宫动脉粗如食指。

（4）阴道和阴门。妊娠初期，阴门收缩紧闭，阴道干涩。妊娠末期，阴唇、阴道因水肿而柔软，利于胎儿产出。

三、妊娠期

妊娠期是母畜妊娠全过程所经历的时间，妊娠期长短主要受畜种、品种、年龄、胎儿、环境条件等因素的影响。各种动物的平均妊娠期见表 5-1。

flow-like with no obvious interval, which is called pregnancy pulse. This is one of the characteristics of pregnancy. The intensity and appearing times of pregnancy pulses of different female animals are different. At the end of pregnancy, the uterine artery of cattle and horses is as thick as the index finger.

(4) Vagina and vulvae

At the begin of pregnancy, the vulvae contracts tightly and the vagina is dry. At the end of pregnancy, the labium and vagina are soft due to edema, which is conducive to fetal delivery.

3 Gestational period

Gestation period is the whole process of maternal pregnancy. The length of gestation period is mainly affected by species, breed, age, fetus, environmental conditions and other factors. The table below shows the average gestational periods of various animals (Table 5-1).

表 5-1　各种动物的妊娠期（d）
Table 5-1　The average gestational periods of various animals (day)

动物 Animals	平均 Average	范围 Range	动物 Animals	平均 Average	范围 Range
牛 Cattle	282	276~290	马 Horse	340	320~350
水牛 Buffalo	307	295~315	驴 Donkey	360	350~370
牦牛 Yak	255	226~289	骆驼 Camel	389	370~390
猪 Pig	114	102~140	犬 Dog	62	59~65
绵羊 Sheep	150	146~161	猫 Cat	58	55~60
山羊 Goat	152	146~161	兔 Rabbit	30	28~33

任务 1　牛的妊娠诊断
Task 1　Pregnancy Diagnosis of Cattle

任务描述

准确地对配种后的母牛进行妊娠诊断，特别是早期诊断，对提高母牛的受胎率有十分重要的意义。母牛的妊娠诊

Task Description

Accurate pregnancy diagnosis, especially early diagnosis, is very important to improve the conception rate of cattle. There are many methods to diagnose preg-

断方法有多种，如外部观察法、超声波诊断法、直肠检查法、阴道检查法及实验室诊断法等。请选择适合的方法诊断母牛是否妊娠，如已妊娠请判断妊娠的阶段，并推算其预产期。

nancy in cattle, such as visual observation, ultrasonic diagnosis, rectal palpation, vaginal examination and laboratory diagnosis. Choose the appropriate method to diagnose whether the cow is pregnant or not. If the cow is pregnant, please judge the stage of pregnancy and calculate the expected date of delivery.

任务实施

一、外部观察法

（一）准备工作
母牛 10～20 头、听诊器。

（二）诊断方法

1. 视诊

询问、调查母牛的生产阶段和发情周期情况；观察母牛的采食情况、膘情、行为、性情及体形变化；观察母牛的外生殖器官变化和乳房变化。

2. 触诊

在母牛妊娠后期，用手掌或拳头在其右膝壁前方、髋部下方推动腹壁感触胎儿的"浮动"以判断妊娠情况和妊娠时间。

3. 听诊

妊娠6个月以后，在安静的场所于母牛右髋部下方或膝壁内侧听取胎心音。

（三）结果判定

1. 视诊

经配种的母牛，过了两个情期仍不出现发情，则可初步确定为已妊娠。母牛妊娠后，食欲增加，膘情好转，被毛光亮，性情温顺，行动谨慎，易离群，怕拥挤。妊娠初期，外阴收缩紧闭，有皱纹。妊娠中后期（5个月以后），腹围、乳房增大，右侧腹壁突出，有的牛

Task Implementation

1 Visual Observation

1.1 Preparations
10-20 cows, stethoscopes.

1.2 Diagnostic methods

1.2.1 Visual inspection

To inquire and investigate the production stages and estrus cycle of cows, to observe the feeding status, fatness, behaviors, temperaments and changes of body shape, to observe the changes of external reproduction organs and breasts.

1.2.2 Palpation

In late pregnancy, the floating fetus can be felt to judge the state and period of pregnancy by pushing the abdominal wall with the palm or fist in front of the right knee wall and below the shank.

1.2.3 Auscultation

After 6 months of pregnancy, fetal heart beats can be heard below the right shank or inside the knee wall in a quiet place.

1.3 Result judgement

1.3.1 Visual inspection

After two estrous periods, if the mated cow still does not show estrus, which can be preliminarily determined as pregnant. After pregnancy, some physiological phenomena will appear in the cow such as increased appetite, improved fatness, good bloom, gentle temperament, cautious action, easy to leave the herd and afraid of crowding. At the beginning of pregnancy, the vulva contracts tightly and wrinkles. In the mid-late pregnancy

项目五 妊娠诊断
Project V PREGNANCY DIAGNOSIS

会出现腹下和四肢水肿。8个月左右，右侧腹壁可见到胎动。此方法在妊娠中后期观察比较准确，但不能在早期做出确切诊断。

2. 触诊

根据胎儿大小和母牛肥胖程度，大约不到5%的母牛在妊娠5个月时就能感触到胎动，10%～50%于妊娠6个月，70%～80%于妊娠7个月，90%以上于妊娠9个月。

3. 听诊

一般胎心音的频率为100次/min以上，为母牛心音频率的2倍以上。

二、直肠检查法

直肠检查是一种隔着直肠壁检查母牛卵巢、子宫及孕体状况的妊娠诊断方法，该方法可以判断母牛是否妊娠以及妊娠所处的阶段。直肠检查法是母牛早期妊娠诊断最常用和最可靠的方法之一。可在配种后40～60 d判断是否妊娠，准确率达90%以上。

（一）准备工作

母牛站立保定，将尾巴拉向一侧，掏出宿粪，清洗外阴。检查人员将指甲剪短、磨光，穿好工作服，戴上长臂手套，清洗并涂抹润滑剂。

（二）诊断方法

检查人员站在母牛正后方，将尾巴拉向一侧，手指并拢呈锥形缓慢旋转伸入直肠，摸到子宫颈，然后手掌展平向前滑动，分别触摸两个子宫角或孕体状

(after 5 months), the abdominal circumference and breast enlarge, the right abdominal wall protrudes, and some cows have edema of the hypogastrium and limbs. At about 8 months, fetal movement can be observed in the right abdominal wall. This method is more accurate in the mid-late pregnancy, but can not be used to make a definite diagnosis in the early stage.

1.3.2 Palpation

According to the size of fetus and degree of obesity of cows, fetal movement can be felt in less than 5% of the cows at 5 months of gestation, 10%-50% at 6 months, 70%-80% at 7 months and 90% at 9 months.

1.3.3 Auscultation

In general, the frequency of fetal heart beats is more than 100 times per minute, which is more than twice the number of cows.

2 Rectal palpation

Rectal palpation is a method of pregnancy diagnosis that examines the ovary, uterus and embryo of cows across the rectal wall. This method can determine whether the cows are pregnant or not and the stage of pregnancy. Rectal palpation is one of the most common and reliable methods for early pregnancy diagnosis in cows. Pregnancy can be judged 40-60 days after mating, and the accuracy rate is more than 90%.

2.1 Preparations

The cow is fixed standing and the tail is pulled to one side to remove the feces in the rectum and clean the vulva. Inspectors should cut and polish their nails, put on work clothes, put on long arm gloves and apply lubricants.

2.2 Diagnostic methods

The inspector stands behind the cow, pulls the tail to one side, slowly rotates his fingers together into the rectum in a conical shape, touches the cervix, then slides the palm forward and touches the two uterine

况。若不确定母牛是否妊娠，可继续向前，在子宫角尖端外侧或下侧寻找卵巢，触摸卵巢有无黄体存在。

检查项目主要包括：卵巢的位置、大小和有无黄体存在；子宫角的质地、状态、大小和位置；胚泡的大小和位置；子叶是否出现及其大小；子宫动脉的粗细和有无妊娠脉搏等。

（三）结果判定

1. 未妊娠特征

青年母牛的子宫及卵巢均位于骨盆腔内。经产多次的牛，生殖器官比较大，子宫角位于骨盆入口前缘的腹腔内。两子宫角大小相等，形状相似，弯曲如绵羊角状，经产牛有时右角略大于左角，弛缓、肥厚。能够清楚地摸到子宫角间沟，子宫质地良好，有弹性和收缩性。卵巢大小及形状视有无黄体或较大的卵泡而定。

2. 妊娠特征

（1）妊娠 18～25 d。子宫角变化不明显，一侧卵巢上有黄体存在，可初步诊断为妊娠。

（2）妊娠 30 d。两侧子宫角不对称，孕角比空角粗大、松软，有波动感，收缩反应不明显，空角较厚且有弹性（图 5-6）。

（3）妊娠 45～60 d。子宫角和卵巢垂入腹腔，孕角比空角约粗两倍，壁薄柔软，波动明显（图 5-7）。用手指从子宫角尖端向基部轻轻滑动，可

horns or the embryo. If you are not sure the cow is pregnant, you can move your hand further forward to look for the ovary at the outside or below the tip of uterine horn and feel the presence or absence of the corpus luteum.

Examination items include: the location, size and presence of corpus luteum; the texture, state, size and location of uterine horn; the size and location of blastocyst; cotyledons and its size; the size and pulse of uterine artery.

2.3 Result judgement

2.3.1 Unpregnant characteristics

The uterus and ovaries of young cows are located in the pelvic cavity. In the cows with multiple births, the reproductive organs are larger, and uterine horns are in the abdominal cavity at the front edge of the pelvic entrance. The two uterine horns are equal in size and similar in shape, curved like sheep horns. Sometimes the right uterine horn is slightly larger than the left one, which is relaxed and hypertrophic. The sulcus between uterine horns can be clearly touched, the uterus is good, elastic and contractile. The size and shape of ovaries depend on the presence or absence of corpus luteum or larger follicles.

2.3.2 Pregnant characteristics

(1) 18th to 25th day of gestation

The change of uterine horn is not obvious. There is corpus luteum on one side of the ovary, which could be initially diagnosed as pregnancy.

(2) 30th day of gestation

The uterine horns are asymmetric. The gestational uterine horn is thicker and softer than the empty uterine horn. It has a sense of fluctuation and the contraction reaction is not obvious. The empty uterine horn is thicker and elastic (Figure 5-6).

(3) 45th to 60th day of gestation

The uterine horns and ovaries fall into the abdominal cavity. The gestational uterine horn is about twice as thick as that of the empty uterine horn, the wall is

感到有胎囊滑过，胎儿大如鸭蛋或鹅蛋。角间沟稍平坦，但仍能分辨。此时一般可确诊。

thin and soft with obvious fluctuations (Figure 5-7). Sliding gently from the tip of the uterine horn to the base with your finger, you can feel the fetal sac slipping through. The fetus is as big as duck's or goose's egg. The sulcus between uterine horns is slightly flat, but it can still be distinguished. At this time, the diagnosis can generally be confirmed.

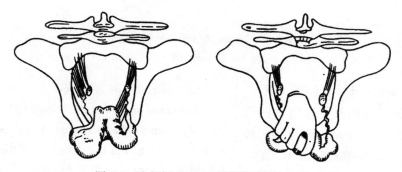

图 5-6　牛妊娠 30 d 子宫形状及触诊方法

Figure 5-6　Uterine shape and palpation method of 30th day of gestation in cattle

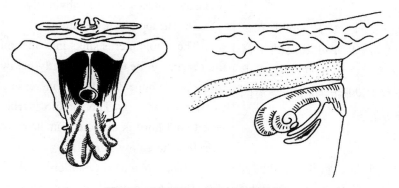

图 5-7　牛妊娠 60 d 子宫形状及位置

Figure 5-7　Uterine shape and location of 60th day of gestation in cattle

（4）妊娠 90 d。角间沟消失，子宫颈移至耻骨前缘（图 5-8）。孕角大如婴儿头，波动明显，空角比平时增大一倍，子叶如蚕豆大小。

(4) 90th day of gestation

The sulcus between uterine horns will disappear and the cervix will move to the front of the pubic bone (Figure 5-8). The gestational uterine horn is as big as the baby's head, and the fluctuation is obvious. The empty uterine horn is twice as big as usual, and the cotyledons are as big as broad beans.

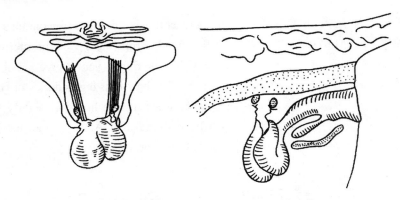

图 5-8 牛妊娠 90 d 子宫形状及位置

Figure 5-8 Uterine shape and location of 90th day of gestation in cattle

(5) 妊娠 120 d。子宫沉入腹底，只能触摸到子宫后部及子宫壁上的子叶，子叶直径为 2~5 cm。子宫颈沉移至耻骨前缘下方，不易摸到胎儿。子宫中动脉逐渐变得粗如手指，并出现明显的妊娠脉搏。

(6) 妊娠 5 个月。子宫全部沉入腹腔，在耻骨前缘稍下方可以摸到子宫颈。子叶逐渐增大，大小如胡桃或鸡蛋。孕角侧子宫中动脉已明显，空角侧尚无或稍有妊娠脉搏。

(7) 妊娠 6 个月。胎儿已经很大，子宫沉至腹腔底。胎盘突如鸽蛋大小，孕角侧子宫中动脉粗大，孕脉亦明显。

(8) 妊娠 7 个月。由于胎儿更大，故从此以后都容易摸到。两侧子宫中动脉均有明显的孕脉，但空角侧较弱。

(9) 妊娠 8 个月。子宫颈回到骨盆

(5) 120th day of gestation

The uterus sinks into the bottom of the abdomen. We can only touch the the back of uterus and the cotyledons on the wall of uterus, and the diameter of the cotyledons is 2-5 cm. The cervix sinks beneath the front edge of the pubic bone, it is difficult to touch the fetus. The middle uterine artery gradually becomes thicker as a finger, and pregnancy pulses become obviously.

(6) 5th month of gestation

The whole uterus sinks into abdominal cavity, and the cervix can be touched slightly below the front edge of the pubic bone. Cotyledons increase gradually as big as walnuts or eggs. The middle uterine artery on the gestational horn is obvious, but there was no or slight pulse on the empty horn.

(7) 6th month of gestation

The fetus is very large and the uterus sinks into the bottom of the abdominal cavity. The size of placenta protuberant is like a pigeon's egg. The middle uterine artery on the side of gestational horn is thick, and the pregnancy pulses are also obvious.

(8) 7th month of gestation

Because the fetus is bigger, so it is easy to be touched thereafter. Both sides of the middle uterine artery have obvious pregnancy pulses, but the empty horn side is weak.

(9) 8th month of gestation

前缘或骨盆腔内，很容易触及胎儿，胎盘突大如鸭蛋，两侧子宫中动脉孕脉显著，孕角侧子宫后动脉的孕脉也已清楚，个别牛即使到产前也不显著。

（10）妊娠9个月。胎儿的前置部分进入骨盆入口，所有的子宫动脉均有显著孕脉，手伸入肛门，只要贴在骨盆侧壁上即可感到孕脉颤动。

（四）注意事项

在直肠检查过程中，检查人员应小心谨慎，避免粗暴。如遇母牛努责，应暂时停止操作，等待直肠收缩缓解时再进行检查。母牛做早期妊娠检查时，要抓住典型征状，根据子宫角和卵巢的变化，做出综合判断。以下几种情况要注意区分：

（1）一些母牛配种后20d已妊娠，但偶尔出现假发情现象，直肠检查妊娠征状不明显，对这种牛应慎重对待，无成熟卵泡者不应配种。

（2）怀双胎母牛，妊娠2个月时两侧子宫角是对称的，不能依其对称而判为未孕。

（3）正确区分妊娠子宫和子宫疾病。妊娠90~120 d的子宫容易与子宫积液、积脓混淆。子宫积液或积脓时，一侧子宫角及子宫体膨大，子宫有不同程度的下沉，但子宫并无妊娠征状，也无子叶出现。

项目五 妊娠诊断

Project V Pregnancy Diagnosis

The cervix returns to the pelvic front or pelvic cavity, and it is easy to touch the fetus. The placenta protuberant is as big as duck's egg. The pregnancy pulses of the middle uterine artery on both sides are significant. And the pulses of the posterior uterine artery on the gestational horn are also clear. Even before delivery, the pregnancy pulses of some individual cattle are not significant.

(10) 9th month of gestation

The anterior part of the fetus enters the pelvic entrance, and all uterine arteries have significant pregnancy pulses. The pregnancy pulses can be felt when sliding the hand into the anus and attachs it to the pelvic lateral wall.

2.4 Cautions

In the process of rectal palpation, the inspectors should be careful and mild. In case of cow abdomen contracting, the operation should be suspended until rectal contraction is relieved. When cows are examined for early pregnancy, we should grasp the typical symptoms and make a comprehensive judgement according to the changes of uterine horns and ovaries. The following situations should be distinguished：

(1) Some pregnant cows occasionally appear false estrus phenomenon 20 days after mating, and pregnancy symptoms are not obvious by rectal palpation. This kind of cows should be treated carefully and should not be mated without mature follicle.

(2) Cows are pregnant with twins, the uterine horns are symmetrical at 2nd month of gestation and can not be judged infertile according to their symmetry.

(3) Correctly distingwish between pregnancy uterus and uterine diseases. The uterus of 90th-120th day of gestation is easily confused with effusion and pyometra. When uterine effusion or pyometra occurs, the horns and bodies of the uterus on one side are enlarged and the uterus sinks to varying degrees. However, the uterus has no pregnancy signs and cotyledons.

(4) Correct distinction between pregnant uterus and urinary bladder. The uterus at 60th-90th day of gestation may be confused with urine-filled bladder, especially at 2nd month of gestation. The bladder is clear, without involvement on both sides, but the cattle have a sulcus at the front of the uterus and a cervix at the back. The enlarged bladder surface is not smooth and has a reticular feeling. The surface of the fetus is smooth and is of even texture.

(5) Correct distinction between pregnancy pulses and fibrillating pulses. The uterine artery of the nonpregnant cattle with a history of multiple delivery also has tremor similar to that of the pregnancy pulses, which should be paid attention to.

3 Ultrasound diagnosis

Ultrasound diagnosis can detect the pregnancy of cows using the physical characteristics of ultrasound. At present, portable B-mode ultrasonograph (Figure 5-9) is widely used. It can diagnose early pregnancy of cows, with simple operation and high accuracy.

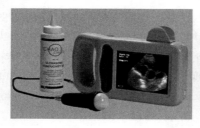 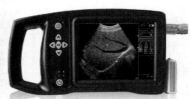

Figure 5-9　Veterinary B-mode ultrasonograph

3.1　Preparations

The cow is fixed standing and the tail is pulled to one side to remove the feces in the rectum. Inspectors cut and polish their nails, put on work clothes, roll up sleeves, put on disposable gloves and apply lubricants. Connect the water-proof probe to the machine, turn on the power and strap B-mode ultrasonograph onto your arm.

（二）检查方法

牛的B超仪诊断法包括体外探查法、阴道探查法和直肠探查法，其中直肠探查法应用效果最好。将涂抹耦合剂的探头慢慢送入受检母牛直肠，隔着直肠壁紧贴子宫角缓慢移动并不断调整探查角度，观察B超实时图像，直至出现满意图像为止，按冻结键，根据图像判断是否妊娠（图5-10）。探头离子宫越近，探测的图像就越清楚。

3.2 Diagnostic methods

B-mode ultrasound diagnosis in cattle includes in vitro exploration, vaginal exploration and rectal exploration, among which rectal exploration is the best one. The probe smeared with the couplant is slowly fed into the rectum of the cow, moved slowly across the rectal wall close to the uterine horns to adjust the exploration angle. The real-time image of B-mode ultrasonography is observed until satisfactory images appear. Pregnancy is judged according to the image by pressing the freezing key(Figure 5-10). The closer the probe is from the uterus, the clearer the image will be.

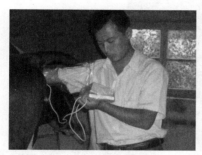
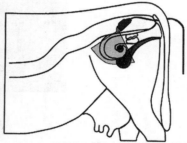

图 5-10　牛 B 超妊娠检查

Figure 5-10　B-mode ultrasound pregnancy examination in cattle

（三）结果判定

在子宫内检测到胚囊、胚斑和胎心搏动即判为阳性；声像图显示一个或多个圆形液性暗区，判为可疑；声像图显示子宫壁无明显增厚变化、无回声暗区，判为阴性。

1. 空怀母牛子宫声像图

子宫体轮廓清晰，内部呈均匀的等强度回声，子宫壁很薄（图 5-11）。

2. 妊娠母牛子宫声像图

妊娠母牛的子宫壁增厚，配种后 12~14d 子宫腔内出现不连续、无反射小区，即为聚有液体的胚泡。妊娠 20d 胚泡结构中出现短直线状的胚体。妊娠

3.3 Result judgement

Embryo sac, plaque and fetal heart beat are detected in uterus, which could be judged as positive. The sonogram shows one or more circular liquid dark areas, which is suspected. The sonogram shows that there is no obvious change in uterine wall and no echo dark areas, which is negative.

3.3.1 Uterine sonogram of nonpregnant cow

The uterus has a clear outline, uniform internal echo of equal intensity, and a thin uterine wall(Figure 5-11).

3.3.2 Uterine sonogram of pregnant cow

The uterine wall of pregnant cows become thicker, discontinuous non-reflex zones appear in uterine cavity 12th to 14th day after mating, they are liquid-filled blastocysts. A short and straight linear embryo appears in the blastocyst at 20th day of gesta-

22d 可探测到胚体心跳。妊娠 22～30d，胚体呈 C 形（图 5-12）。33～36d，呈现清晰的胚囊和胚斑图像，胚囊如一指大小，胚斑如 1/3 指大小（图 5-13），子宫壁结构完整，边界清晰，胚囊液性暗区大而明显。40d 以上，胚囊和胚斑均明显可见，有时还可见胎心搏动（图 5-14）。

tion. The heartbeat of embryo could be detected at 22nd day of gestation. The shape of embryo is like letter C at 22nd-30th day of gestation (Figure 5-12). The embryo sac and plaque could be clearly displayed at 33rd-36th day of gestation, the embryo sac is the size of one finger, and the plaque is the size of one third finger (Figure 5-13). The structure of uterine wall is complete, the boundary is clear, and the liquid dark area of embryo sac is large and obvious. Over 40 days, the embryo sac and plaque are clear, and sometimes fetal heartbeat is also visible (Figure 5-14).

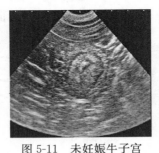

图 5-11　未妊娠牛子宫

Figure 5-11　Uterus of nonpregnant cow

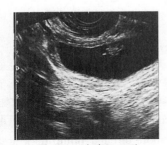 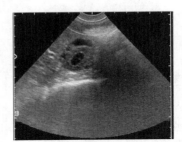

图 5-12　妊娠 27 天　　　　　　　　　图 5-13　妊娠 33 天

Figure 5-12　27th day of gestation　　　Figure 5-13　33rd day of gestation

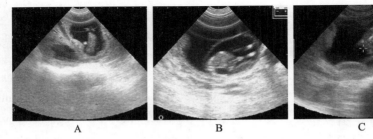

图 5-14　妊娠 40d 以上

Figure 5-14　Pregnancy over 40 days

A. 妊娠第 45 天　B. 妊娠第 66 天　C. 妊娠第 89 天

A. 45th day of gestation　B. 66th day of gestation　C. 89th day of gestation

项目五 妊娠诊断
Project V Pregnancy Diagnosis

（四）注意事项

（1）液性暗区是圆形或近圆形强回声区（一般为纯黑色），在暗区周围有一圈完整的灰色环状组织。

（2）暗区一定位于子宫角内部。部分假象图是一些挤压空腔或者血管形成，位于子宫角外面，周围没有完整灰色环状组织。探查时还有可能误将黄体或大卵泡当作胚囊。

（3）当牛患有子宫炎或子宫积液等疾病时，图像也会出现许多不规则黑色暗区或暗区内有光斑（暗区也位于子宫角内），从而导致误诊。

3.4 Cautions

(1) The liquid dark area is a circular or near-circular strong echo area(usually pure black), surrounded by a circle of intact gray annular tissue.

(2) Dark areas must be located inside the uterine horns. Some dark areas show squeezing cavity or blood vessels, located outside the uterine horns, and there is no intact gray ring around them. It is also possible to mistake corpus luteum or large follicles as embryo sacs during exploration.

(3) When cows suffer from hysteritis or uterine hydrops, there are also many irregular black areas or dark areas with spots. The dark areas are also located in the uterine horn, which sometimes leads to misdiagnosis.

任务 2 羊的妊娠诊断
Task 2 Pregnancy Diagnosis of Sheep

羊的妊娠诊断
Pregnancy diagnosis of sheep

任务描述

随着养羊企业的快速发展，需要大量繁殖优质羔羊扩大生产规模。因此，母羊配种后应尽早进行妊娠诊断，及时做好空怀母羊的补配工作。如何快速辨别母羊是否妊娠？

Task Description

With the rapid development of sheep breeding enterprises, a large number of high-quality lambs are required to expand production scale. Therefore, the pregnancy diagnosis should be carried out as soon as possible after mating and do the breeding work of the non-pregnant ewes in time. How to quickly identify whether the ewe is pregnant or not?

任务实施

一、外部观察法

1. 准备工作

将母羊放于运动场内，让其自由活动。

2. 诊断方法

询问、调查母羊的生产阶段和发情周期情况；观察母羊的采食情况、膘

Task Implementation

1 Visual observation

1.1 Preparations

Put the ewes in the playground and let them move freely.

1.2 Diagnostic methods

To investigate the production stage and estrus cycle of ewes. To observe the feeding status, fatness, behavior,

情、行为、性情及体形变化；观察母羊的外生殖器官变化和乳房变化。

3. 结果判定

母羊妊娠后，发情周期停止，食欲增加，毛色光亮，性情温顺，行动谨慎小心，好静、喜卧。妊娠初期，外阴收缩紧闭，有皱纹。妊娠 2～3 个月，腹围增大，右后腹部突出，乳房增大。

二、触诊法

（一）一般触诊法

检查人员面对母羊头部，双腿夹持母羊颈部，两手以抬抱的方式在母羊腹壁前后滑动，触摸是否有胎泡。

（二）直肠-腹壁触诊法

1. 诊断方法

将待查母羊用肥皂灌洗直肠排出粪便，使其仰卧，将涂抹润滑剂的触诊棒（直径 1.5cm、长 50cm）贴近脊椎插入直肠约 30cm。一只手用触诊棒把直肠轻轻挑起以便托起胎泡，另一只手在腹壁上触摸。

2. 结果判定

触诊时，如有胎泡即表明已妊娠；如果摸到触诊棒，将棒稍微移动位置，反复挑起触摸 2～3 次，仍摸到触诊棒即表明未妊娠。使用该方法时，动作要小心、轻缓，以防损伤直肠和胎儿。该方法准确率很高，在早期妊娠诊断非常重要。

三、超声波探测法

（一）准备工作

1. 保定母羊

用 B 超诊断仪探查早期妊娠时，母羊侧卧、仰卧或站立均可。如大群检

temperament and changes of body shape. To observe the changes of external reproduction organs and breasts.

1.3 Result judgement

After pregnancy, the estrus cycle of ewes stops, some physiological phenomena appear, such as increased appetite, good bloom, gentle temperament, cautious action. At the beginning of pregnancy, the vulva contracts tightly and wrinkles. At 2nd-3rd month of gestation, the abdominal circumference and breast enlarge, the right abdominal wall protrudes.

2 Palpation

2.1 General palpation

The inspector holds the ewe's neck with both legs facing the head, and slide his hands around the abdominal wall to determine whether there are embryos.

2.2 Palpation of rectum-abdominal wall

2.2.1 Diagnostic methods

The ewe undergoes a rectal lavage with soap water to excrete feces from the rectum and made to lie on its backs. The palpation stick (1.5 cm in diameter and 50 cm in length) with lubricant is inserted into the rectum close to the spine for about 30 cm. One hand gently lifts the rectum with a palpation stick to hold up the fetus, and the other hand feels the abdominal wall.

2.2.2 Verdict-making

When palpating, if there is a fetus, it indicates pregnant. If touching the palpation stick, you need to move the stick slightly, repeatedly lift and touch 2-3 times, it indicates unpregnant if still touching the palpation stick. We should be careful and gentle to prevent damage to the rectum and fetus. The accuracy of this method is very high, and it is very important in early pregnancy diagnosis.

3 Ultrasound diagnosis

3.1 Preparations

3.1.1 Fixed ewes

B-mode ultrasonograph can detect early pregnancy of ewes by lying on the side or back or standing. For

查，采用保定架保定；如少量检查，可由助手扶持，使母羊安静站立。如需进行腹侧检查，则需在右腹侧后部剪毛。

2. B超仪准备

连接好B超仪探头，打开B超仪，调节好对比度、辉度和增益，使其适合当时当地的光线强弱及检测者的视觉。准备好耦合剂。

（二）诊断方法

1. 直肠探查

应用于妊娠早期（妊娠40 d以内），将探头插入直肠内约15 cm，越过膀胱，向两侧转45°角进行扫查，以探测到胎水或子叶为判定妊娠阳性依据。

2. 腹部探查

检查者蹲于羊体一侧，将涂抹耦合剂的探头紧贴母羊腹部皮肤，朝盆腔入口方向定点做扇形扫查，如探查到胎儿（胎头、胎心、脊椎或胎蹄）可判定为阳性。妊娠早期在乳房两侧及膝皱襞之间无毛区域，或两乳房的间隔处进行探查；妊娠中后期可在右侧腹壁进行探查。

（三）结果判定

1. 未妊娠母羊

子宫一般位于膀胱前方或前下方，为近圆形的弱反射，直径在10.0mm以上。有时断面中央可见一个很小的暗区，直径为2.0~3.0mm，但不随配种天数增加而变大，子宫可随膀胱积尿的程度而上下移位。

mass inspection, the sheep should be fixed by a frame. For individual inspection, the ewe can be supported by an assistant to make them stand quietly. If ventral examination is needed, the wool of back part of right ventral need to be shorn.

3.1.2 Preparation of B-mode ultrasonograph

Connect the probe, turn on the B-mode ultrasonograph, adjust the contrast, brightness and gain so as to make it suitable for the local light intensity and the inspector's vision at that time. Prepare the couplant.

3.2 Diagnostic methods

3.2.1 Rectal exploration

Rectal exploration is applied in the early pregnancy (within 40 days). The probe is inserted into the rectum about 15 cm deep, crossed the bladder and turned 45 degrees to both sides for scanning. Pregnancy can be determined if fetal fluid or cotyledons are detected.

3.2.2 Abdominal exploration

The examiners squat on one side of the sheep, press the probe with the coupling agent close to the abdominal skin, and make a fixed-point sector scanning towards the pelvic entrance. If the fetus (head, heart, spine or hoof) is detected, it can be judged to be pregnant. In the early pregnancy, the exploration site is located at hairless area between the breast and the knee fold, or the interval between the two breasts. In the middle and late pregnancy, the exploration site is located at the right abdominal wall.

3.3 Result judgement

3.3.1 Unpregnant ewes

The uterus is usually located in front of or below the bladder. It is a nearly circular weak reflex with a diameter of more than 10.0 mm. Sometimes a small dark area with a diameter of 2.0-3.0 mm can be seen in the center of the section, but it does not become larger as time passes after mating. The uterus can move up and down with the variations of bladder urine amout.

2. 妊娠母羊

子宫内出现暗区（胚囊），最初为单个小暗区。配种后18～20d，可探查到胚斑，为椭圆形的弱反射光斑，长约6mm，位于子宫暗区的下方或一侧。妊娠23d时，膀胱下面出现带状的无回声区（胎水）。妊娠25d时，呈不规则的无回声区，可见到胎体，但看不到胎心搏动。妊娠30d时，不规则无回声区扩大并移至膀胱前下方，无回声区边缘或其中有"扣状"小回声，为胎盘，羊膜囊内有胎体，细心观察可见到胎心搏动。诊断早孕的准确率为97%左右。

3.3.2 Pregnant ewes

Initially a small dark area(embryo sac) appears in the uterus. Embryo spot can be detected 18-20 days after mating. It is an elliptical weak reflection spot with about 6 millimeters long, located below or on one side of the dark area of uterus. On the 23rd day of gestation, a banded anechoic zone (fetal fluid) appears below the bladder. On the 25th day of gestation, there is an irregular anechoic zone in which the fetal body could be seen, but no fetal heart beat. On the 30th day of gestation, the irregular anechoic area is enlarged and moves to the anterior and inferior part of the bladder. The edge of the anechoic area or with the small "button" echo in it is the placenta, and the fetal heart beat can be observed carefully in the amniotic sac. The accuracy of early pregnancy diagnosis is about 97%.

任务3　猪的妊娠诊断
Task 3　Pregnancy Diagnosis of Pigs

任务描述

母猪配种后，应尽早检出空怀母猪，及时补配，防止空怀。这对于保胎、缩短胎次间隔、提高繁殖力和经济效益具有重要的意义。母猪妊娠诊断常用的方法有外部观察法、返情检查法、超声波诊断法等。规模化猪场多采用B超检测，也有用A超检测。二者有哪些异同点？

Task Description

After mating, the nonpregnant sows should be detected as soon as possible, and be mated again in time. This is of great significance to the preservation of fetus, shortening the interval of parities, improving the fecundity and economic benefits. The commonly used methods of sow pregnancy diagnosis include visual observation, re-estrus inspection and ultrasound diagnosis. Pigs in large farms are mostly detected by B-mode ultrasonograph, together with some using A-mode ultrasonograph. What are the similarities and differences between them?

任务实施

一、外部观察法

1. 准备工作

母猪10～20头，自由活动。

Task Implementation

1　Visual observation

1.1　Preparations

To prepare 10-20 sows and let them move freely.

项目五　妊娠诊断
Project V　Pregnancy Diagnosis

2. 诊断方法

询问、调查母猪的生产阶段和发情周期情况；观察母猪的采食情况、膘情、行为、性情及体形变化；观察母猪的外生殖器官变化和乳房变化。

3. 结果判定

母猪妊娠后，表现为发情周期停止，食欲增强，膘情好转，被毛光亮，性情温顺，行动小心。妊娠初期，外阴苍白、皱缩，妊娠中后期（2个月后），腹围增大、隆起。

二、返情检查

妊娠诊断最普通的方法是根据配种后17～24 d是否恢复发情。母猪的发情周期一般为21 d左右，正常情况下，母猪配种后20多天不再出现发情表现，可初步确定妊娠，第二个情期仍不发情，即可确定妊娠。个别母猪妊娠后，偶尔也会有发情表现，因此，最好与超声波诊断结合使用。

三、超声波诊断法

（一）A超诊断仪诊断母猪妊娠

1. 准备工作

待检母猪、A超诊断仪（图5-15）、耦合剂。

2. 诊断方法

将待测母猪站立保定，在右腹侧最后一对乳头上方处，涂抹适量耦合剂，将探头紧贴皮肤对子宫进行扇形扫描。当发出稳定声音后，记录结果。然后，在左侧再检测一次，以验证结果。

1.2　Diagnostic methods

To investigate the production stage and estrus cycle of sows. To observe the feeding status, fatness, behavior, temperament and changes of body shape. To observe the changes of external reproduction organs and breasts.

1.3　Result judgement

After pregnancy, the estrus cycle of sows stops, some physiological phenomena appear such as increased appetite and fatness, good bloom, gentle temperament, acting cautiously. At the beginning of pregnancy, the vulva is pale and wrinkled. The abdominal circumference is enlarged and bulged in the middle and later period of pregnancy(2 months later).

2　Re-estrus inspection

The most common method of pregnancy diagnosis is based on whether estrus is restored 17-24 days after mating. The estrus cycle of sows is generally about 21 days. Normally, pregnancy is preliminarily determined if the sow does not show estrus for 20 days after mating. Pregnancy can be determined if the second cycle is still not estrus. Some sows will occasionally have estrus behaviour after pregnancy, so it is best to diagnose pregnancy in combination with ultrasonic.

3　Ultrasonic diagnosis

3.1　Pregnant diagnosis of sows by A-mode ultrasonograph

3.1.1　Preparations

Sows, A-mode ultrasonograph(Figure 5-15), coupling agent.

3.1.2　Diagnostic methods

The sow is fixed standing. Smear some coupling agent on the top area of the last pair of nipples on the right ventral side, then attach the probe to the uterus for sector scanning. When a steady sound is emitted, the results are recorded. Then, check again on the left to verify the results.

图 5-15 猪用 A 超诊断仪
Figure 5-15 A-mode ultrasonograph for swine

3. 结果判定

当听到连续的"嘀嘀"声，可诊断为妊娠；当听到断续的"嘀嘀"声，多次调整探头方向，仍无连续响声，则诊断为未妊娠。

（二）B 超诊断仪诊断母猪妊娠

1. 准备工作

妊娠 18～35 d 的母猪数头，B 超诊断仪，耦合剂。

2. 诊断方法

用湿毛巾擦除母猪腹部污物。母猪站立时在探头上涂抹耦合剂，侧卧或趴卧时在探查部位涂抹耦合剂，将探头指向耻骨前部和骨盆入口方向。随妊娠天数的增加，探查部位逐渐前移。

3. 结果判定

（1）妊娠母猪（图 5-16）。当看到典型的孕囊暗区即可确认早孕阳性。妊娠早期（妊娠 18～21 d）子宫中出现孕囊，呈直径约 1 cm 的圆形暗区，通常为一个或 2～3 个相邻的暗区，位于膀胱暗区的前下方。随妊娠期的延长，暗区不断扩大且呈多个不规则圆形、椭圆形暗区。妊娠 26 d 后，胚胎逐渐显出

3.1.3 Result judgement

Pregnancy can be diagnosed when the device emits continuous beep sound. When hearing intermittent beep sound, adjusting the direction many times, if there is still no continuous sound, unpregnancy can be diagnosed.

3.2 Pregnant diagnosis of sows by B-mode ultrasonograph

3.2.1 Preparations

Some sows at 18-35 days of gestation, B-mode ultrasonograph, coupling agent.

3.2.2 Diagnostic methods

Wipe the sow's abdomen clean with a wet towel. When the sow is standing, the probe is coated with coupling agent. While lying on the side or on the abdomen, the exploratory site is coated with coupling agent. Place the probe in the direction of the anterior part of the pubic bone and the pelvic entrance. As pregnancy progresses, the site is moved forward gradually.

3.2.3 Result judgement

(1) Pregnant sows (Figure 5-16)

When the typical dark area of the gestational sac appears, the positive early pregnancy can be confirmed. In early pregnancy (18th-21st day), there are gestational sacs in the uterus, usually one or 2-3 adjacent dark areas about 1 cm in diameter, located in the front and bottom of the bladder. As pregnancy progresses, dark areas continue to expand, showing irregular circular and oval shapes. After 26 days of

胎儿固有轮廓，胎头、躯体及四肢逐渐发育完善，出现胎动及内脏器官（肝、胃等）。

（2）空怀母猪。空怀母猪子宫显示为不规则圆形的弱反射区。一周后应及时复查，以免误诊。

gestation, the embryo gradually shows the inherent contour, the fetal head, body and limbs are well-developed, and fetal movement and visceral organs (liver, stomach, etc.) appear.

(2) Nonpregnant sows

The uterus of nonpregnant sows shows a weak reflex area of irregular circular shape. A week later, another examination should be made to avoid misdiagnosis.

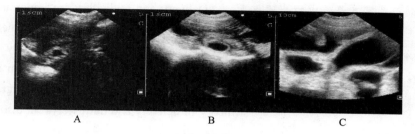

图 5-16　猪 B 超妊娠检测声像

Figure 5-16　Sonography of pregnant diagnosis in sows by B-mode ultrasonograph

A. 妊娠 18d　B. 妊娠 21d　C. 妊娠 28d

A. 18th day of gestation　B. 21st day of gestation　C. 28th day of gestation

猪 B 超诊断

Pregnant diagnosis of sows by B-mode ultrasonograph

项目六　接产与助产
Project Ⅵ　Delivery and Midwifery

项目导学

接产和助产是家畜繁殖中的一项重要工作，助产不当或不及时助产直接关系到母仔生命的安危及产后疾病的预防。因此，生产中既要掌握母畜接产与助产的基本理论，还要能熟练进行接产与助产操作。

Project Guidance

Delivery and midwifery are important tasks in livestock reproduction. Improper or untimely midwifery may have a direct impact on the safety of newborns and their mothers as well as postpartum diseases prevention. Therefore, we should not only grasp the basic theory of delivery and midwifery, but also be skilled in operation.

学习目标

>>> 知识目标

• 熟悉母畜的分娩预兆和分娩过程，为判断是否正常分娩奠定基础。

• 了解难产的原因和种类。
• 熟悉难产救助的基本原则。

• 理解分娩机理及影响分娩过程的因素。

>>> 技能目标

• 能准确判断母畜的分娩预兆及难产迹象。
• 能熟练实施母畜分娩前准备及助产技术。
• 能正确掌握难产检查及处理技术。
• 会正确护理新生仔畜。

Learning Objectives

>>> Knowledge Objectives

• Be familiar with the parturition signs and parturition process, which provides the basis for correctly diagnosing whether the parturition is normal or not.
• Know the causes and kinds of dystocia.
• Be familiar with the rules of rescue when dystocia occurs.
• Understand the mechanism and influencing factors of parturition process.

>>> Skill Objectives

• Judge the parturition sings and dystocia signs accurately.
• Master the prenatal preparation work and the skills of midwifery.
• Master the examination and treatment skills of dystocia correctly.
• Nurse newborn animals correctly.

一、分娩机理

分娩是指雌性动物经过一定的妊娠期，胎儿在母体内发育成熟，母体将胎儿及其附属物从子宫内排出体外的生理过程。引起分娩启动的因素是多方面的。分娩是由激素、神经和机械等多种因素的协同、配合，母体和胎儿共同参与完成的。

1. 机械刺激

妊娠末期，由于胎儿生长很快，胎水增多，胎儿运动增强，使子宫不断扩张，承受的压力逐渐升高。当子宫的压力与子宫肌高度伸张状态达到一定程度时，便可引起神经反射性子宫收缩和子宫颈的舒张，从而导致分娩。

2. 母体激素的变化

临近分娩时，母体内孕激素分泌减少或消失，雌激素、前列腺素（$PGF_{2\alpha}$）、催产素分泌增加，同时卵巢及胎盘分泌的松弛素促使产道松弛。母体在这些激素的共同作用下发生分娩。分娩前后相关激素的变化趋势见图6-1。

1 Mechanism of parturition

Parturition is defined as the physiologic birth process by which the pregnant uterus delivers the fetus and placenta from the maternal organs. Parturition involes both the matrixes and the fetus, and it is completed by a complex interaction of many factors, such as hormonal, neural, mechanical factors, etc.

1.1 Mechanical stimulation

The fetus grow fast, motion enhancement and the fetal water increase induce the pressure on the uterus distending and expand at the final stage of gestation. When the pressure on uterus and the stretched condition of myometrium reach up to a point, it will cause nerve reflex contraction of uterus and diastole of cervix, which then leads to parturition.

1.2 Hormonal changes in dams

Near parturition, the levels of progesterone fall or disappear, while the levels of estrogen, prostaglandin ($PGF_{2\alpha}$) and oxytocin increase, at the same time relaxin secreted by ovary and placenta, gives the birth canal laxity. The variation curve of related hormones before and after parturition is shown in Figure 6-1.

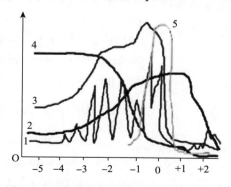

图 6-1 母畜分娩前后体内激素水平的变化

Figure 6-1 Hormonal changes in dams before and after parturition

1. $PGF_{2\alpha}$ 2. 雌激素 3. 胎儿肾上腺皮质激素 4. 孕激素 5. 催产素

1. prostaglandin ($PGF_{2\alpha}$) 2. estrogen 3. fetal adrenocorticotropic hormone 4. progesterone 5. oxytocin

3. 神经系统

神经系统对分娩并不是完全必需的，但对分娩过程具有调节作用。如胎儿的前置部分对子宫颈和阴道产生刺激，通过神经传导使垂体后叶释放催产素导致分娩。此外，多数母畜在夜间分娩，可能是由于外界光线及干扰减少，黑暗和安静的环境易于接受来自子宫及软产道的神经刺激。

4. 胎儿因素

胎儿发育成熟后，脑垂体分泌促肾上腺皮质激素，促使胎儿肾上腺分泌肾上腺皮质激素。胎儿肾上腺皮质激素引起胎盘分泌大量雌激素及母体子宫分泌大量前列腺素，并使孕激素水平下降。雌激素使子宫肌对各种刺激更加敏感，而且还能促使母体自身释放催产素。所以在母体的催产素与前列腺素的协同作用下，激发子宫收缩，导致胎儿娩出。

5. 免疫学机理

妊娠后期，胎儿发育成熟时，胎盘发生脂肪变性，胎盘屏障受到破坏，胎儿和母体之间的联系中断，胎儿被母体免疫系统识别为"异物"而排出体外。

二、 分娩预兆

母畜在分娩前，在生理和形态上都会发生一系列变化，通常将这些变化称为分娩预兆。根据乳房、外阴部、骨盆等变化，可预测母畜的分娩时间，以便做好产前准备，确保母仔平安。

1.3 Nervous system

The nervous system is not absolutely necessary for parturition, but it regulates the process of parturition. For example, the fore-lying part of fetus can stimulate cervix and vagina, leading to posterior pituitary release oxytocin by nerve conduction, then cause labour. Otherwise, most dams labour at night probably because of the reduction of ambient light and perturbation making them easy to accept nerve stimulation from uterus and soft birth canal in dark and quiet environment.

1.4 Fetal influence

After the fetus matures, its pituitary gland secretes adrenocorticotropic hormone, induces adrenal gland to secrete adrenal hormone. Fetal contrical hormone not only causes placenta and maternal uterus to secrete large amounts of estrogen and prostaglandin, but also decreases the level of progestin. Estrogen will make the myometrium more sensitive to various stimuli and prompt the matrixes to release oxytocin. The synergistic effect of the oxytocin and prostaglandin induces the uterine contraction, then leads to the delivery of the fetus.

1.5 Immunological mechanism

When the fetus matures, it has been shown to induce placental steatosis, while the placenta barrier is destroyed and the connection between the fetus and the matrixes is interrupted at the final stage of gestation. Finally, the fetus is recognized as a foreign-body by the maternal immune system and thus being "expelled" from the body.

2 Signs of parturition

A series of physiological and morphological changes in prenatal dams are usually called signs of parturition. We can predict the parturition time of dams according to the changes of breast, vulva and pelvis, in order to make prenatal preparation and ensure the safety of both mothers and offspring.

分娩前乳房迅速发育，膨胀增大，有时还出现浮肿，个别有漏乳现象。牛的乳房变化比其他家畜明显。在分娩前数天到一周左右，阴唇逐渐变松软、肿胀、体积增大，阴唇皮肤上的皱褶展平，并充血稍变红。从阴道流出的黏液由浓稠变稀薄，尤以牛和羊最为明显。骨盆韧带从分娩前1～2周开始软化，到分娩前12～36 h，荐坐韧带后缘变得非常松软，外形消失，尾根两侧下陷，只能摸到一堆松软组织，即"塌窝"，但初产牛这些变化不明显。

分娩前精神状态变化比较明显，母畜出现食欲不振、精神沉郁、徘徊不安和离群寻找安静地等现象。猪在临产前6～12 h，出现衔草做窝现象。家兔有扯咬胸部被毛和衔草做窝现象。马和驴在临产前数小时，表现不安、频繁举尾、蹄踢下腹部和时常起卧及回顾腹部等。

三、分娩过程

（一）影响分娩过程的因素

1. 产力

将胎儿从子宫中排出的力量称为产力，包括阵缩和努责。阵缩是指子宫肌有节律的收缩，贯穿于整个分娩过程，是分娩的主要动力。努责是指腹肌和膈肌的强有力收缩，是胎儿产出的辅助动力。

2. 产道

产道是分娩时胎儿由子宫排出体外的通道，可分为软产道和硬产道两部分。

Obvious enlargement of the mammary gland occurs before parturition, as well as swollen, and sometimes edema or milk leakage occurs, especially in cows. From a few days to a week before parturition, the labia become pliable, edematous, bulky gradually and slightly reddish with congestion, the wrinkles on the skin flatten out. Mucus from the vagina becomes thinner, especially in the cow and the ewe. The pelvic ligament becomes relaxed and flaccid about 1-2 weeks before parturition. Around 12-36 hours before parturition, the posterior border of the sacrosciatic ligament becomes more flaccid and the root of tail sags bilaterally, while only a pile of soft tissues can be touched. But these changes are not obvious in primiparous cows.

Before parturition, the mental state of the dam will change significantly, with symptoms such as anorexia, depression, restlessness and seeking isolation to seek quiet place. Swines build nests about 6-12 hours before parturition, rabbits bite the chest coat and make nests. Horses and donkeys show restlessness, frequent tail lifting, kicking the lower abdomen, frequent sitting-up and lying-down and looking back at the abdomen in few hours before parturition.

3 Parturition process

3.1 Factors affecting the parturition process

3.1.1 Force of delivery

Force of delivery means the force that expels the fetus from the uterus, including contracture and straining. Contracture means the rhythmic myometrial contractions, which is the main power of parturition and runs through the whole process of parturition. Straining refers to the strong contraction of abdominal and diaphragm muscles, which is the auxiliary power of labor.

3.1.2 Birth canal

The birth canal is the channel through which the fetus is expelled from the uterus, it can be divided into two parts: soft birth canal and hard birth canal.

(1) 软产道。包括子宫颈、阴道、前庭和阴门。分娩时子宫颈逐渐松弛直至完全开张，阴道、阴道前庭和阴门也能充分松软扩张。

(2) 硬产道。就是骨盆，可分为四个部分：①入口，即骨盆的腹腔面，入口大而倾斜，形状圆而宽阔，胎儿则容易通过。②骨盆腔，即骨盆入口至出口之间的空间。③出口，即骨盆腔向臀部的开口。④骨盆轴，代表胎儿通过骨盆腔时所走的路线，骨盆轴越短越直，胎儿通过越容易。各种母畜的骨盆特点如表 6-1 和图 6-2 所示。

3. 胎儿与母体的关系

分娩时，胎儿与母体产道的相互关系，对胎儿产出有很大影响。此外，胎儿的大小和是否畸形也影响胎儿能否顺利产出。

(1) Soft birth canal

The soft birth canal includes cervix, vagina, vestibule and vulva. During parturition, the cervix gradually relaxes until it is fully opened, and the vagina, vestibule and vulva are also fully relaxed and dilated.

(2) Hard birth canal

The hard birth canal is the pelvis, which can be divided into four parts: ①The entrance is the abdominal side of the pelvis. The entrance to the pelvis is large and oblique, round and wide in shape, which is easy for the fetus to pass through. ②The pelvic cavity refers to the space between the entrance and exit of the pelvis. ③The outlet means opening of the pelvic cavity to the buttocks. ④The pelvic axis is the route through which the fetus passes through the pelvic cavity. The shorter and straighter the pelvic axis is, the easier for the fetus to pass through. The pelvic characteristics of various dams are shown in table 6-1 and Figure 6-2.

3.1.3 The relationship between fetus and dam

During parturition, whether the fetus can be delivered smoothly is influenced by the relationship between the fetus and the maternal birth canal, the size of the fetus and whether it is malformed.

表 6-1 各种母畜骨盆特点
Table 6-1 Pelvic characteristics of dams

	牛 Cattle	马 Horse	猪 Swine	羊 Sheep
入口 Entrance	竖长椭圆 Vertical ellipse	圆形 Circular	近乎圆形 Almost circular	椭圆形 Ellipse
出口 Outlet	较小 Small	大 Large	很大 Great	大 Large
倾斜度 Inclination degree	较小 Small	大 Large	很大 Great	很大 Great
骨盆轴 Pelvic axis	曲线形 Curvilinear	浅弧形 Shallow arc	较直 Relatively straight	弧形 Arc
分娩难易程度 Difficulty of parturition	较难 Relatively difficult	易 Easy	很易 Very easy	易 Easy

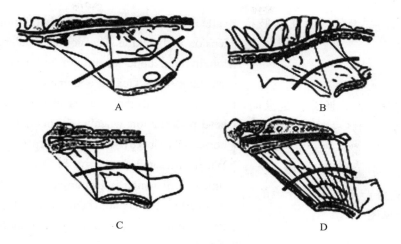

图 6-2　各种母畜的骨盆轴
Figure 6-1　Pelvic axis of dams
A. 牛　B. 马　C. 猪　D. 羊
A. cow　B. horse　C. sow　D. ewe

（1）胎向。指胎儿纵轴与母体纵轴的相互关系，分为纵向、竖向和横向。纵向是正常胎向，横向和竖向都属反常胎向，易发生难产。

（2）胎位。指胎儿背部与母体背部的关系，分为上位、下位和侧位。上位是正常的，下位和侧位是反常的。如果侧位倾斜不大，仍可视为正常。

（3）前置。又称先露，是指胎儿先进入产道的部位。头和前肢先进入产道为头前置（正生）；臀部和后肢先进入产道为臀前置（倒生）。

（4）胎势。指胎儿在母体内的姿势，一般分为伸展或屈曲的姿势。正常

(1) Fetal orientation

Fetal orientation refers to the relationship between longitudinal axis of the fetus and that of matrixes, which can be divided into longitudinal, vertical and transverse. Normal fetal orientation is longitudinal, the others are abnormal orientations and prone to causing dystocia.

(2) Fetal position

Fetal position refers to the relationship between the dorsum of the fetus and that of matrixes, which can be divided into upper, inferior and lateral position. The upper position is normal, the inferior and lateral positions are abnormal. It is still regarded as normal if inclination angle of the lateral position is not obvious.

(3) Presentation

Presentation means that the part of fetus enters the birth canal first. If the head and forelimbs first enter the birth canal, it is called anterior presentation, while if the buttocks and hind limbs first enter the birth canal, it is called posterior presentation.

(4) Fetal posture

The posture of the fetus in the maternal body is

的胎势为头纵向、上位、胎儿前肢抱头、后肢踢腹。

一般母畜分娩时，胎儿多是纵向，头部前置，马占 98%～99%，牛约占 95%，羊为 70%、猪为 54%。牛、羊怀双胎时多为一个正生、一个倒生，猪往往是正倒交替产出。正常分娩的胎位、胎势变化见图 6-3。

called fetal posture, which is generally divided into extension and flexion. The normal fetal posture is head longitudinal, upper position, forelimbs embracing head and hind limbs kicking abdomen.

In general, the fetus is usually longitudinal with head presentation when in labor (about 98%-99% in horse, 95% in cow, 70% in sheep and 54% in sow). When cattle and sheep give birth to twins, most of the time one of which is in normal position and the other is inverted. Piglets are often delivered in alternate pattern of one in normal position and another in inverted position. Fetal position and fetal posture of normal parturition are shown in Figure 6-3.

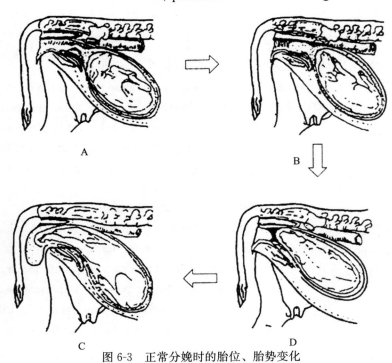

图 6-3　正常分娩时的胎位、胎势变化
Figure 6-3　Fetal position and fetal posture of normal parturition
A. 纵向下位　B. 头、前肢后伸　C. 纵向侧位　D. 纵向上位
A. longitudinal and inferior position　B. head and forelimbs rearward extension
C. longitudinal and lateral position　D. longitudinal and upper position

（二）分娩过程

分娩过程是从子宫肌和腹肌出现阵缩开始，至胎儿及其附属物排出为止。分娩是有机联系的完整过程，按照产道

3.2 Parturition process

Parturition starts with the myometrial and abdominal muscular contractions, until the fetus and its appendages are expelled. Parturition is a complete process,

暂时性的形态变化和子宫内容物的排出情况，习惯上将分娩过程分为子宫颈开口期、胎儿产出期和胎衣排出期3个阶段。

1. 子宫颈开口期

从子宫出现阵缩开始，至子宫颈口完全开张为止。这一期只有阵缩而无努责。初产母畜表现不安、时起时卧、徘徊运动、频频举尾。但经产母畜一般比较安静。

2. 胎儿产出期

从子宫颈口完全开张至排出胎儿为止，阵缩和努责共同作用，而努责是排出胎儿的主要力量。当羊膜随着胎儿进入骨盆入口，便引起膈肌和腹肌反射性和随意性收缩。胎儿最宽部分的排出时间最长，特别是头部通过骨盆及其出口时，母畜努责最强烈。

牛的产出期约为6h，有的长达12h；马、驴产出期约为12h，有的长达24h；猪为3~4h；绵羊为4~5h；山羊为6~7h。

3. 胎衣排出期

从胎儿排出后到胎衣完全排出为止。胎儿排出后，一方面母体胎盘的血液循环减弱，子宫黏膜腺窝的紧张性降低；另一方面胎儿胎盘的血液循环停止，绒毛膜上的绒毛体积缩小，间隙增大，使绒毛很容易从腺窝中脱落。

胎衣排出得快慢，因各种家畜的胎盘组织结构不同而有差异。弥散型胎盘组织结合比较疏松，胎衣容易脱落，所以排出最快，猪、马和驴的胎衣排出就

which can be divided into three stages: dilation of the cervix, expulsion of the fetus and expulsion of the placenta, according to the temporary morphological changes of the birth canal and the expulsion of uterine contents.

3.2.1 Dilation of the cervix

This stage begins with contraction of the uterus and ends with fully opened cervical orifice, only contracture without straining exists in this stage. The dam of first parturition is restless, having frequent sitting-ups and lying-downs, raising tails frequently. But the multiparous dams are usually quiet.

3.2.2 Expulsion of the fetus

This stage begins with fully opened cervical orifice and ends with the expulsion of the fetus, by the combined effect of contracture and straining, but straining is the main force to expel the fetus. When the fetus enclosed in the amnion enters the pelvic entrance, it causes contraction of the diaphragm and abdominal muscles. The widest part of the fetus takes the longest time to be expelled, as well as the strongest straining, especially when the head of the fetus passes through the pelvis and its outlet.

The duration of expulsion in cow, horse, donkey, sow, ewe and goat is respectively about 6-12, 12-24, 3-4, 4-5 and 6-7 hours.

3.2.3 Expulsion of the placenta

This stage begins with the expulsion of the fetus and ends with complete expulsion of placenta. After the fetus is expelled, the blood circulation of the maternal placenta and the tension of the uterine mucosal glandular fossa decrease, on the other hand, the blood circulation of the fetal placenta stops, and the volume of villi on the chorion shrinks and the gap increases, which makes the villi easily fall off from the glandular fossa.

The rate of placental expulsion varies with the structure of placenta in different livestock. The placentas of swine, horse and donkey belong to diffused-type placenta that the tissue structure is loosened and the placenta is easy to fall off. The expulsion period of

属于这种情况，猪的胎衣排出期为10～60 min，马和驴为5～90 min。牛、羊的胎盘属于子叶型，结合比较紧密，子宫肌收缩时不容易影响腺窝，只有当母体胎盘组织的张力减小时，胎儿胎盘的绒毛才能脱落下来，所以需要时间较长。牛的胎衣排出期为2～8 h，绵羊为0.5～4 h，山羊为0.5～2 h。

四、难产

（一）难产分类

在分娩过程中，如果母畜产程过长或胎儿排不出体外，称为难产。由于发生的原因不同，可将难产分为产力性难产、产道性难产和胎儿性难产3种。

（1）产力性难产。子宫阵缩及努责微弱、阵缩及破水过早和子宫疝气等引起的产力不足而导致的难产。

（2）产道性难产。子宫捻转，子宫颈、阴道及骨盆狭窄、产道肿瘤等引起的难产。

（3）胎儿性难产。胎儿过大，胎位、胎向和胎势不正等引起的难产。

在上述三种难产中，以胎儿性难产最为多见，在牛的难产中约占75%，在马、驴的难产中可达80%。

（二）难产的预防

（1）切忌母畜配种过早，若母畜尚未发育成熟，分娩时容易因骨盆狭窄造成难产。

（2）妊娠期间，要合理饲养母畜，给予营养全面的饲料，以保证胎儿发育

placenta is 10-60 minutes in swine and 5-90 minutes in horse and donkey. The placentas of cattle and sheep belong to cotyledon type that the structure is compact. The myometrial contractions of this placental type does not tend to affect the glandular fossa, that the villi of the placenta can fall off only when the tension of the maternal placenta tissue is reduced, so it takes a long time to expel the placenta. The expulsion periods of placenta in cattle, sheep and goats are 2-8 hours, 0.5-4 hours and 0.5-2 hours, respectively.

4 Dystocia

4.1 Classification of dystocia

Dystocia can be defined as the inability of the dam to expel neonates through the birth canal from the uterus, or prolonged labor that beyond the normal parturition time. According to the different causes, dystocia can be divided into three types: dystocia of labor force, dystocia of birth canal and fetal dystocia.

(1) Dystocia of labor force is caused by a deficiency of expulsive forces, such as weakness of contracture and straining, uterine hernia, ect.

(2) Dystocia of birth canal is caused by uterine torsion, birth canal tumor and stenosis of the cervix, vagina and pelvis.

(3) Fetal dystocia is caused by excessive fetal size and erroneous of fetal orientation, position and posture.

Among the three types of dystocia mentioned above, fetal dystocia is the most common, accounting for about 75% in cases of dystocia in cow and 80% in horse and donkey.

4.2 Prevention of dystocia

(1) The dam should not be prematurely bred. If the dam has not yet developed maturely, it is susceptible to dystocia due to pelvic stenosis during parturition.

(2) During gestation, the dam should be fed reasonably and given perfect nutrition to ensure the development of the fetus and the health of the dam and re-

和母畜健康，减少难产发生的可能性。妊娠末期，要适当减少蛋白质饲料，以免胎儿过大。

（3）安排适当的使役和运动，可提高母畜对营养物质的利用，使全身及子宫肌的紧张性提高，有利于分娩时胎儿的转位，防止胎衣不下及子宫复位不全等。

（4）做好临产检查，对分娩正常与否做出早期诊断。

duce the possibility of dystocia. At the end of pregnancy, reduce protein feed to avoid excessive fetal growth.

(3) By taking on proper tasks and exercises, the utilization of nutrients can be improved, and the tension of the whole body and uterine muscle can be improved, which is beneficial to the translocation of the fetus during delivery, and can prevent retention of the placenta and incomplete uterine reduction etc.

(4) Early diagnosis of whether the dam can undertake normal labour is made by prenatal examination.

任务 1 牛的接产与助产
Task 1 Delivery and Midwifery of Cattle

任务描述

王华大学毕业后到一家大型奶牛场工作，新员工岗前培训结束后，王华被分配到产房工作。他如何胜任产房的工作呢？

Task Description

After graduation, Wang Hua worked in a large dairy farm. He was assigned to the parturition room after the pre-job training. How could he be qualified for the parturition room?

任务实施

Task Implementation

一、产前的准备

1. 产房

产房的基本要求为宽敞明亮、清洁干燥、通风透光、没有贼风、保温良好和环境安静。使用前应进行全面的清扫消毒，把产房的墙壁和地面、运动场、饲槽、分娩栏等打扫干净。

2. 器械及物品

准备好接产用具和药品，如水盆、肥皂、纱布、药棉、剪刀、助产绳、催产药和消毒药（如碘酒、酒精等），有

1 Prenatal preparation

1.1 Parturition room

The basic requirements of parturition room are being spacious and bright, clean and dry, well-ventilated and transparent, no wayward wind blowing in, good heat preservation and quiet. The walls and floor of parturition room should be thoroughly cleaned and disinfected, the playground, as well as the trough and the delivery stable before being used.

1.2 Instruments and articles

Prepare the delivery utensils and medicines, such as water basin, soap, gauze, medicinal cotton, scissors, midwifery rope, oxytocic and disinfectant(such as iodine

条件的牛场最好准备一套产科器械。

3. 助产人员

助产人员应具有一定的助产经验，随时观察和检查母牛的健康状况，严格遵守接产操作程序。另外，由于母牛多在夜间分娩，所以要做好夜间值班。

4. 母牛

预产期前1～2周将待产母牛转入产房。对临产前的母牛，用温水清洗外阴部，并用煤酚皂液或高锰酸钾溶液彻底消毒。分娩时，让母牛左侧卧或站立，以免胎儿受瘤胃压迫产出困难。

二、助产方法

母牛正常分娩时，一般无须人为干预，接产人员的主要任务是监视分娩状况，并护理好新生犊牛，清除犊牛鼻腔、口腔内黏液，剪断脐带，擦干皮肤，及时喂初乳。只有确定母牛发生难产时再进行助产。

（1）在胎儿进入产出期时，应及时确定胎向、胎位、胎势是否正常。正常分娩的姿势见图6-4：一种是正生上位，头颈和两前肢伸直，且头颈在两前肢的上面；另一种是倒生上位，两后肢伸直，胎儿以楔状进入产道。如果胎向、胎位及胎势均正常，不必急于将胎儿拉出，待其自然娩出。如果胎势异常，可将胎儿推回子宫进行整复矫正。出现倒生时，应迅速拉出胎儿，免得胎儿腹部进入产道后，脐带可能被压在骨盆底下，造成窒息死亡。

tincture, alcohol). It is best to prepare a set of obstetric equipment for cattle farms with bigger budgets.

1.3 Midwife

Midwives should be experienced, strictly follow the procedures of delivery and be able to observe and check the health status of the cow at any time. In addition, the midwife should carry out night duties accordingly, as the cow usually give birth at night.

1.4 Cow

Transfer the predelivery cows to the parturition room 1-2 weeks before the expected parturition date, wash the vulva of the cow thoroughly with warm water and disinfect with lysol or potassium permanganate. When labouring, keep it in standing or left-lateral position to avoid fetal dystocia due to the pressure of rumen.

2 Midwifery methods

Human intervention is not needed when the cow labours normally. The main task of the midwife is to monitor the delivery situation, nurse the newborn calves, wipe the mucus from nasal cavity and mouth, cut the umbilical cord, dry the skin and feed colostrum in time. The midwifery is needed only when dystocia happens.

（1）When the fetus enters the delivery period, it should be determined whether the fetal orientation, position and posture are normal or not. The normal posture of the fetus during parturition is shown in Figure 6-4. One is head presentation, with the head and neck on the two forelimbs and stretch straightly, the other is inverted presentation, with the two hind limbs straight and the fetus entering the birth canal in a wedge-shape. If the fetal orientation, position and posture are normal, there is no need to pull the fetus out, just waiting for its natural delivery, but if those are abnormal, the fetus can be pushed back to the uterus for correction. When the fetus is in inverted presentation, the midwife should pulled out the fetus quickly to prevent the asphyxia and death, because the umbilical cord may be pressed under the pelvis after the abdomen enters the birth canal.

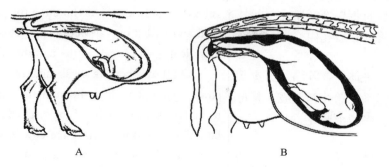

图 6-4 正常分娩姿势

Figure 6-4　Normal posture of the fetus during parturition
A. 正生上位　B. 倒生上位
A. head presentation and upper position　B. foot presentation and upper position

（2）当胎儿头部已露出阴门外，如果排出时间过长，也应助产。如果胎膜尚未破裂（图 6-5），应及时撕破。拉出胎儿时，要注意保护好母牛的会阴部，以免发生阴道破裂或子宫脱出。如果破水过早，产道干滞，可注入液状石蜡进行润滑。胎儿头部尚未露出阴门外时，不要过早扯破羊膜，以防胎水流失，引起产道干涩。

(2) Midwifery also should be implemented when the fetal head has been exposed outside the vulva for a long time. If the fetal membrane is not ruptured (Figure 6-5), midwives should tear it in time. When pulling out the fetus, attention should be paid to protect the perineum of the cow to avoid vaginal rupture or uterine prolapse. In order to prevent the dry and astringent of birth canal caused by the early loss of fetal water, liquid paraffin could be injected as lubricant, also midwives should not tear the amniotic membrane prematurely if the fetal head has not yet protruded out of the vulva.

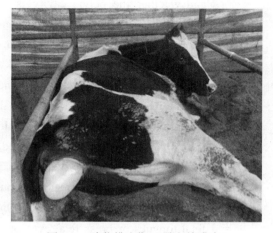

母牛分娩

Parturition of cow

图 6-5 胎儿排出期（露出羊膜囊）

Figure 6-5　Expulsion of the fetus (exposure of amniotic sac)

（3）当胎儿腹部通过阴门时，将手伸至其腹下握住脐带根部和胎儿一起拉出，以免脐血管断在脐孔内。

（4）如果母牛体弱，阵缩、努责无力，可用助产绳系住胎儿两前肢，趁母牛努责时顺势缓慢拉出胎儿。

（5）当母牛站立分娩时，应双手接住胎儿，以免摔伤。

（6）胎衣排出后，及时检查是否完整。如果胎衣排出不完整，说明母体子宫有残留胎衣，要及时处理使胎衣完全排出。胎衣排出后，应立即取走，以免母牛吞食后引起消化紊乱。

三、难产救助方法

（一）难产的检查

1. 产道检查

该项目主要检查产道是否干燥、有无损伤、水肿或狭窄，子宫颈开张程度、有无损伤或瘢痕，骨盆腔是否狭窄及有无畸形、肿瘤等，并要注意产道内液体的颜色及气味。

2. 胎儿检查

常见的胎儿性难产有头颈侧弯、前肢屈曲、后肢屈曲、胎儿过大、胎儿畸形等（图6-6）。另外，还要检查胎儿的死活。正生时，助产人员可将手伸入胎儿口腔，轻拉舌头或轻压眼部或牵扯刺激前肢，注意观察有无生理反应，如口吮吸、舌收缩、眼转动、肢伸缩等；也可触诊颌下动脉或心区有无搏动。倒生时，触诊脐带有无动脉搏动，也可牵拉

(3) When the abdomen of the fetus goes through the vulva, the midwife should hold the root of the umbilical cord and pull it out, as well as the fetus, to prevent the umbilical vessel break in the umbilicus.

(4) If the cow is weak and the contracture and straining are powerless, the midwifery rope can be used to tie the two forelimbs of the fetus, then the fetus will be pulled out slowly while the cow is straining.

(5) When the cows are standing to labour, midwives should catch the fetus with both hands in order to avoid falling.

(6) Check the completeness of the placenta after it is discharged. If the placenta is incomplete, it indicates that there is residual placenta in the uterus of dams. After the placenta is discharged, it should be removed immediately to avoid being swallowed by the cow, which can induce digestive disturbance.

3　Rescue methods for dystocia

3.1　Examination of dystocia

3.1.1　Examination of birth canal

During this examination, the birth canal, cervix and the pelvis should be checked, for example, whether the birth canal is injured, dry, edematous or narrow, whether the cervix is fully dilated, injured or scarred, and whether the pelvic cavity is narrow and abnormal. In the meantime, pay attention to the color and odor of the liquid in the birth canal.

3.1.2　Examination of fetal

Fetal dystocia includes head and neck lateral bending, flexion of forelimb and hind limb, oversize and malformation of fetus (Figure 6-6). In addition, whether the fetus is dead should be examined. When the fetus is head presentation, midwives can palpate the submandibular artery or cardiac region for pulsation and extend their hands into the oral cavity of the fetus, gently pull fetal tongues or press fetal eyes or stimulate the forelimbs and pay attention to the physiological reactions, such as mouth sucking, tongue contraction, eye rotation,

刺激后肢注意有无反射活动,或将食指轻轻伸入肛门检查其收缩反射。

limb contraction, etc. When the fetus is foot presentation, midwives can palpate umbilical cord for arterial pulsation and pay attention to the reflex activities of the fetus by pulling and stimulating hind limb or penetrating the anus with index finger.

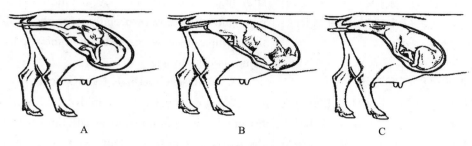

图 6-6 常见胎儿性难产类型

Figure 6-6 Common types of fetal dystocia

A. 头颈左弯 B. 前肢屈曲 C. 髋关节屈曲

A. left bending of head and neck B. forelimb flexion C. hip joint flexion

(二) 难产的救助原则

当发生难产时,应镇静对待,查明原因,及时采取适宜措施。

(1) 救助目的要明确,尽可能保证母仔双方平安。在没有可能的情况下,一般优先考虑母畜安全,但在胚胎移植时,则首先保证良种犊牛的存活。

(2) 助产时,尽量避免产道感染和损伤,注意器械的严格消毒和规范使用。

(3) 母畜横卧分娩时,尽量使胎儿的异常部分向上,便于操作。

(4) 为了便于推回或拉出胎儿,尤其是产道干涩时,应向产道内灌注肥皂水或液状石蜡润滑剂。

(5) 矫正胎儿反常姿势时,尽量将胎儿推回子宫内进行矫正,因为产道空间有限,不易于操作。要掌握好推回胎

3.2 Rescue principles of dystocia

When dystocia occurs, midwives should treat it calmly, find out the cause and take appropriate measures in time.

(1) The purpose of the rescue should be clear and ensure the safety of both the dam and the young as far as possible. In general, the safety of the dam is the priority, but the survival of well-bred calves is the first priority in case of embryo transfer.

(2) During midwifery, midwives should try to avoid infection and injury of the birth canal and pay attention to strict disinfection and standardized use of instruments.

(3) Try to make the abnormal part of the fetus upward so as to easy operation when the dam lay in labor.

(4) In order to push back or pull out the fetus expediently, midwives should inject soapy water or paraffin oil into the birth canal especially when the birth canal is dry and astringent.

(5) When correcting abnormal fetal posture, midwives should try to push the fetus back into the uterus, as the limited space of the birth canal is not easy to op-

儿的时机，尽量选择在阵缩的间歇期进行。矫正困难时，可进行剖宫术或截胎术。

（6）拉出胎儿时，应随着母牛的努责而用力，并注意保护好母牛的会阴部，尤其是初产母牛，胎头通过阴门时，会阴易发生撕裂。

（三）难产的救助方法

当发现难产时，要及时、果断地进行救助。母牛难产时，先注入润滑剂或肥皂水，再将胎儿顺势推回子宫，胎位矫正后，再顺其努责轻轻拉出，严防粗暴硬拉。对于阵缩、努责微弱或子宫颈狭窄的母牛，可注射雌激素和催产素，刺激子宫收缩和宫颈开张。对于胎儿过大或难以矫正、无法助产的母牛，要及时进行剖宫产。

四、新生犊牛护理

1. 清除黏液，注意保温

犊牛刚出生后，立即清除其口腔、鼻腔周围的黏液，尽快擦干犊牛身体上的黏液，以防受凉。如犊牛呼吸困难或停止，迅速抓住犊牛后肢将其倒吊，轻拍犊牛胸部和背部，同时注意清除口鼻倒流出的黏液，直到呼吸顺畅。

2. 断脐消毒

脐带未断裂时，捏住脐带基部，用消毒剪刀距腹部6～8cm处剪断。断端用5%碘酒消毒，一般不结扎，以利于干燥愈合。

erate. The intermittent period of contracture is a good time to push back the fetus. When it is difficult to correct the fetal posture, cesarean section or fetotomy can be performed.

(6) Midwives should pull out the fetus with the straining of the cow and pay attention to protect the perineum of the cow, especially when the fetal head of the primiparous cow passes through the vulva, the perineum is easy to tear.

3.3 Rescue methods for dystocia

When dystocia occures, midwives first injects lubricant or soapy water into uterine lumen and around the fetus, then push the fetus back into the uterus to correct the fetal position, finally gently pull the fetus out with the straining of cow. For cows with weak constrictio, nutria, or narrow cervix, estrogen and oxytocin can be injected to stimulate uterine contraction and cervix opening. For the cows whose fetuses are too large or difficult to correct and unable to accept midwifery, cesarean section should be carried out in time.

4 Nursing of newborn calves

4.1 Clean up mucus and keep warm

Immediately wipe mucus from the newborn's nostrils and mouth with a clean, dry cloth, and gently rub its body to prevent it from catching cold after birth. If the calf has dyspnea or stops breathing, hang the calf by its hind legs immediatly, pat the chest and back of the calf, and pay attention to clean up mucus from the mouth and nose until breathing is smooth.

4.2 Rupture umbilical cord and disinfection

When the umbilical cord is not broken, midwife should pinch its root and cut the umbilical cord 6-8 cm from the calf's body, then dip the umbilical cord in 5% tincture of iodine to prevent infection, which is usually not ligated to facilitate drying and healing.

3. 尽早喂初乳

饲喂初乳的时间越早越好,一般在产后 0.5 h 内,最长不能超过 1 h。初乳饲喂量为犊牛体重的 8%~10%,大型牧场为便于操作一般统一灌服 4 L。

4. 做好记录工作

记录是牛场管理的重点,从新生犊牛开始就要做好相关记录,如犊牛耳号、出生日期、体重,母牛的胎次及产犊难易程度等相关信息。

五、产后母牛的护理

1. 注意胎衣和恶露排出情况

要观察胎衣排出情况,母牛分娩后 8 h 内,胎衣一般可自行排出,若超过 24 h 仍不排出,则称为胎衣不下。根据胎衣滞留时间的长短,可采取药物治疗和手术剥离两种方法。注意恶露的排出情况,恶露一般会持续 10~14 d。恶露排完,说明子宫已复原。如果母牛在产后 3 周仍有恶露排出或恶露腥臭,表示有子宫感染,应及时治疗。

2. 注意产后母牛外阴部和阴道的清洁消毒

对产后 3~5 d 的母牛,应每天用温水、肥皂水、1%~2% 来苏儿或 0.1% 高锰酸钾溶液清洗母牛外阴部并擦干。

3. 加强饲养管理

母牛分娩之后,要及时供给足够的水和麸皮汤或益母草红糖水等,有利于胎衣的排出和子宫的复原。产后 1~2 d 天的母牛还应继续饮用温水,饲喂质量好、易消化的饲料,投料不宜过多,一般 5~6 d 后可以逐渐恢复正常饲养。

4.3 Feed colostrum as early as possible

The calf should be fed colostrum within half an hour after birth, and no more than 1 hour at the latest. Depending on the weight of the calf, you need to feed 8%-10% percent of the calf weight. In large-scale ranch, usually uniformly fed 4L colostrum generally.

4.4 Complete recording task

Recording is the key point of cattle farm management. It is best to record relevant data, such as the ear mark, birth date, weight, gravidity of cow and difficulty levels in calving when they are first born.

5 Nursing of postpartum cow

5.1 Discharge of placenta and lochia

The cow should discharge the placenta within 8 hours after parturition. If the cow has not discharged the placenta within 24 hours after calving, it is called the remained placenta. Medication and surgical stripping can be used according to the length of the stay time of the placenta. The lochia usually lasts for 10-14 days. The exhaustion of lochia indicates that the uterus is restored. If the cow still discharge lochia 3 weeks after parturition or lochia is stinking, it indicates that there is uterine infection and should be treated in time.

5.2 Cleaning and disinfection of vulva and vagina in postpartum cows

The vulva of the cow should be cleaned with warm water, soapy water, 1%-2% lysol or 0.1% potassium permanganate and dried every day within 3-5 days after parturition.

5.3 Strengthen feeding and management

It is necessary to supply sufficient water and bran soup or leonurus brown sugar water in time to facilitate the discharge of placenta and recovery of the uterus in cows after parturition. Cows should continue to drink lukewarm water and to be fed good quality and digestible feed within 1-2 days after parturition. Do not feed too much. Generally, the cows can gradually return to normal feeding 5-6 days after parturition.

任务 2 羊的接产与助产
Task 2　Delivery and Midwifery of Sheep

羊的接产与助产

Delivery and midwifery of ewes

任务描述

做好母羊的接产与助产工作，对于维护母羊健康、提高羔羊成活率均具有重要意义。母羊分娩前有哪些预兆？如何做好新生羔羊的护理工作？

Task Description

It is of great significance to do well in delivery and midwifery of ewes for maintaining the health of ewes and improving survival rate of lamb. What are the signs before parturition? How to nurse newborn lambs?

任务实施

一、产前的准备

1. 产房

产前 3～5d 将产房、运动场、饲槽、分娩栏等清扫干净。为了让分娩母羊熟悉产房环境，在临产前 2～3d 就应将其圈入产房，确定专人管理，随时观察。

2. 接产人员

接产人员应具有丰富的接羔经验，熟悉母羊的分娩规律，严格遵守操作规程。

3. 用具及器械

肥皂、毛巾、药棉、纱布、听诊器、细绳、剪刀、镊子、常用产科器械及必需药品（如酒精、碘酒、新洁尔灭、缩宫素、抗生素等）。

二、助产方法

（1）母羊临产前，首先剪去乳房周围和后肢内侧的毛，用温水洗净乳房，并挤出几滴初乳，再将母羊的尾根、外阴部和肛门洗净，然后用1%来苏儿消毒。

Task Implementation

1　Prenatal preparation

1.1　Parturition room

Parturition room should be cleaned and disinfected 3-5 days before parturition and the ewes should be transferred to the parturition room to familiarize themselves with the environment 2-3 days before parturition. It should be managed by particular person, who can observe and monitor at any time.

1.2　Midwife

Midwives should be experienced, familiar with the rule of ewe delivery, and can strictly follow the operating procedures.

1.3　Instruments and supplies

Soap, towel, cotton, gauze, stethoscope, string, scissors, tweezers, commonly-used obstetric instruments and essential drugs (such as alcohol, iodine, neogeramine, oxytocin, antibiotics, etc).

2　Midwifery methods

（1）Before parturition of ewe, midwives should first shear the wool around the breast and inside the hind limbs, then wash the breasts with warm water and squeeze out a few drops of colostrum, finally wash the tail root, vulva and anus and sterilize them with 1% lysol.

（2）母羊正常分娩时，一般不予干扰，最好让其自行分娩，一般在胎膜破裂、羊水流出后几分钟至30min，羔羊即可产出。正常分娩时，羔羊两前肢夹头先产出，其余随后产下（图6-7）。

(2) Ewes usually deliver lamb within a few minutes to 30 minutes after rupture of membranes and outflow of amniotic fluid. In normal labour, two forelimbs and head come out first and then followed by the rest(Figure 6-7).

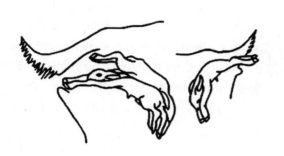

图 6-7　羊的正常胎位

Figure 6-7　normal fetal position of sheep

（3）产双羔时，先后间隔 5～30min，但也偶有长达数小时以上的。因此，当母羊产出第一羔后，如仍表现不安、卧地不起，或起立后又重新躺下、努责等，可用手掌在母羊腹部前方适当用力向上推举，如果可触摸、感觉到光滑的羔体，说明分娩还未结束。

（4）当母羊产道较为狭窄或体乏无力时，需要人工助产。其操作方法：助产人员在母羊体躯后侧，用膝盖轻压其臀部，等羔羊前端露出后，用手推动母羊会阴部，待羔羊头部露出后，然后一手托住头部一手握住前肢，随母羊努责向后下方拉出胎儿。

三、难产救助方法

（1）胎儿过大、阴道狭窄、羊水过早流失等原因形成的难产，先用凡士林

(3)Double lambing is produced at intervals of 5-30 minutes, but sometimes more than a few hours. Therefore, after the ewe produces the first lamb, if the ewe is still restless, lying down, or standing up and lying down alternatively, and straining, etc., midwives can properly put their palms on the abdomen of the ewes and then push them upward. If a smooth lamb body can be touched, it means the parturition is not over yet.

(4)Artificial midwifery is needed when the birth canal of the ewe is too narrow or the ewe is weak. Operation methods: the midwife first stands on the back of ewe's body and gently presses the buttock of the ewe with knees, then pushes the perineum when the front of the lamb is exposed, next holds the head with one hand and holds the forelimbs with the other hand when the head of lamb is exposed, finally pull the fetus out in sync with the straining of the ewe.

3　Rescue methods for dystocia

(1)Dystocia can be caused by oversized fetus, vaginal stenosis and premature loss of amniotic flu-

或液状石蜡润滑阴道，然后把胎儿的两前肢拉出来再送进产道去，反复三四次扩大阴门后，配合母羊阵缩补加外力牵引，帮助胎儿产出。

（2）胎位、胎向、胎势不正时（图6-8），接羔人员应在母羊阵缩间歇时，用手将胎儿轻轻推回腹腔，手也随着伸进阴道，用中指、食指矫正异常的胎位、胎向，并协助将胎儿拉出。

（3）因子宫阵缩及努责微弱引起的产力性难产，可肌内注射或静脉注射适量催产素。

（4）因阴道、阴门狭窄和子宫肿瘤等引起的产道性难产，可在阴门两侧上方将阴唇剪开1~2cm，将阴门翻起同时压迫尾根基部，以使胎头产出而解除难产。如果母羊的子宫颈过于狭窄或不能扩张，应施行剖宫产手术。

（5）双羔同时楔入产道时（图6-9），助产人员应将消毒后的手臂伸入产道将一个胎儿推回子宫内，把另一个胎儿拉出后，再拉出推回的胎儿。如果双羔各将一个肢体伸入产道，形成交叉的情况，则应先辨明关系，顺手触摸肢体与躯干的连接，分清肢体的所属，将其中一个胎儿推回子宫内，把另一个胎儿胎向、胎势调整好后拉出，再拉出推回的胎儿。

四、新生羔羊护理

1. 清除黏液

羔羊出生后，用手先将其口腔、鼻

id. Midwives should lubricate the vagina with vaseline or paraffin oil first, then pull out the fetal forelimbs and send them back to the birth canal. After repeating three or four times, the vulva is expanded, and the fetus will be pulled out with the contracture of the ewe and external traction.

(2) Dystocia can be caused by abnormal fetal orientation, position and posture (Figure 6-8). The midwife should make use of the contracture intervals of ewes to gently push the fetus back to the abdominal cavity with hands and correct the abnormal fetal position with middle finger and index finger, then pull the fetus out.

(3) Labour force dystocia caused by weakness of uterine contracture and straining can be rescued by injecting oxytocin intramuscularly or intravenously.

(4) Birth canal dystocia, which caused by stenosis of vagina and vulva or uterine tumors, can be relieved by cutting the labia 1-2 cm above both sides of the vulva, turning the vulva upside down and pressing the base of the tail root at the same time, so as to deliver the fetal head and resolve dystocia. If the cervix of the ewe is too narrow or unable to dilate, the cesarean section should be performed.

(5) When the twin lambs enter the birth canal at the same time (Figure 6-9), the midwife should extend the sterilized arm into the birth canal, push one fetus back into the uterus, pull one fetus out and then pull the other fetus out. If the twin lambs each extends a limb into the birth canal to form a crossover position, the midwife should first touch the connection between the limb and the trunk, distinguish the ownership of the limb. Then push one fetus back into the uterus, adjust the fetal position of the other correctly and pull it out, then the first one.

4 Nursing of newborn lambs

4.1 Clean up the mucus

After the lamb is born, the mucus in its mouth and

腔里的黏液掏出擦净，以免因呼吸困难、吞食羊水而引起窒息或异物性肺炎。其余部位的黏液让母羊舔干，有利于母羊认羔。如天气寒冷，要用干净布迅速将羔羊身体擦干，以免受凉。

nasal cavity should be wiped out to avoid asphyxia or foreign body pneumonia caused by dyspnea or swallowing amniotic fluid. It is beneficial for ewes to recognize lambs that the ewes lick off the remaining mucus. If the weather is cold, midwives should dry the lamb quickly with a clean cloth, so as to not catch a cold.

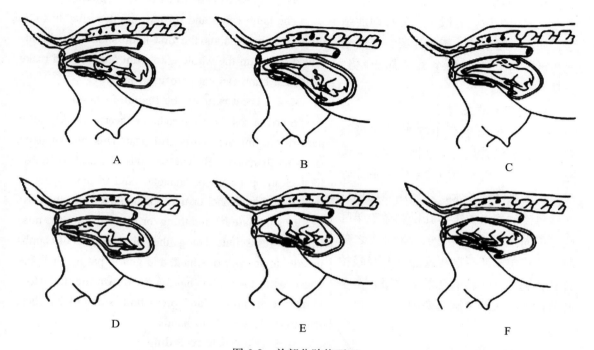

图 6-8 羊部分胎势不正

Figure 6-8 Abnormal fetal position in sheep

A. 头向后仰　B. 头向下弯　C. 头颈侧弯　D. 前肢弯曲　E. 坐骨前置　F. 跗部前置

A. head backward　B. head down bend　C. head and neck lateral bend

D. forelimb curvature　E. ischium presentation　F. tarsal presentation

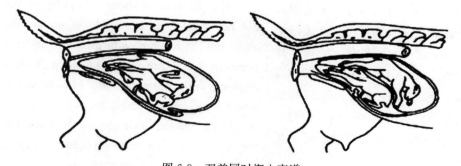

图 6-9 双羔同时楔入产道

Figure 6-3 Twin lambs wedge into the birth canal

2. 断脐

羔羊出生后脐带可自行断裂，或在脐带停止波动后距腹部4～6cm处用手拧断，断端用3％～5％碘酒消毒。

3. 尽快吃初乳

羔羊出生后，一般十几分钟即能站立，如果一胎多羔，不能让第一个羔羊把初乳吃净，应使每个羔羊都能吃到初乳。

4. 假死羔羊处理

有些羔羊出生后不呼吸或呼吸微弱，但心脏仍有跳动，这种现象称为假死。假死羔羊的抢救方法：①先清除呼吸道内的黏液或羊水，然后用酒精棉球或微量碘酒滴入羔羊鼻孔以刺激羔羊呼吸，或向羔羊鼻孔吹气、喷烟使之苏醒；②将羔羊两后肢提起悬空并轻轻拍打其背胸部；③将假死羔羊放平，两手有节律地推压羔羊胸部两侧。

5. 做好记录

将羔羊编号，育种羔羊称量出生重，按要求填写羔羊出生登记表。

五、产后母羊的护理

注意观察胎衣排出情况，羊的胎衣通常在分娩后2～4h内排出。产羔母羊要保暖防潮，产后1h左右，给母羊饮水，一般为1～1.5 L，水温25～30℃，忌饮冷水，可加少许食盐、红糖和麦麸。剪去母羊乳房周围的长毛，用温毛巾擦洗乳房，并挤掉少量乳汁，帮助羔羊吃上初乳。

4.2 Rupture umbilical cord

The umbilical cord can rupture automatically, or it can be broken off by hand 4-6 cm away from the abdomen after the umbilical cord stops fluctuating, and the broken ends should be disinfected with 3%-5% iodine tincture.

4.3 Feed colostrum as early as possible

The lamb can stand up ten minutes after birth. If the ewe is polytocous, the first lamb should not be allowed to eat up the whole colostrum reserve, and make sure each lamb can eat colostrum fairly.

4.4 Treatment of the thanatoid lamb

Some lambs do not breathe or breathe weakly after parturition, but the hearts still beat. This phenomenon is called thanatosis. Rescue methods of thanatoid lambs: First, clean up mucus or amniotic fluid in the respiratory tract, then drip alcohol or trace iodine liquor into the nostrils to stimulate its breathing, or blow air into the nostrils to resuscitate. Hung the hind limbs of lambs upside down with one hand and gently pat its back and chest with the other hand. Lay down the thanatoid lamb flat on ground and press both sides of its chest rhythmically with both hands.

4.5 Complete recording

Midwives should number the lamb, weigh the birth weight of the breeding lamb, and fill in the birth registration form as required.

5 Nursing of postpartum ewes

The placenta of sheep is usually expelled within 2-4 hours after parturition. Ewes should drink water about one hour after parturition, the water volume is generally 1-1.5 litre and the temperature is 25-30℃, no drinking cold water, a little salt, brown sugar and wheat bran can be added to the water. The midwife should cut the long hair around the ewe's udder, scrub the udder with a warm towel, squeeze out a little milk, and help the lamb to eat colostrum.

项目六　接产与助产
Project VI Delivery and Midwifery

任务 3　猪的接产与助产
Task 3　Delivery and Midwifery of Pigs

母猪的接产与助产
Delivery and midwifery of pigs

任务描述 | Task Description

母猪分娩是养猪生产中最繁忙、最细致、最易出问题的环节。为了保证母猪安全分娩和提高仔猪成活率，需熟练掌握接产与助产技术。对于假死仔猪，该如何抢救？

Sow parturition is the busiest, most meticulous and most problematic link in pig production. In order to ensure safe delivery and improve the survival rate of piglets, it is necessary to master the techniques of delivery and midwifery. How to treat the thanatoid piglets?

任务实施 | Task Implementation

一、产前的准备
1　Prenatal preparation

1. 产房

分娩前 5～7d 准备好产房，产房要求温暖干燥、清洁卫生、舒适安静、阳光充足、空气清新，并用 2%～5% 来苏儿或 2%～3% 氢氧化钠溶液喷雾消毒。产前一周将母猪转移到产房中。

1.1　Parturition room

The preparation requirements of the parturition room is similar to those of cattle and sheep, which should be warm and dry, clean and sanitary, comfortable and quiet, sunny and fresh. The sows are transferred to the parturition room one week before parturition.

2. 接产用具

接产时应准备好下列用具：产仔哺育记录卡、手术剪、毛巾、碘酊、高锰酸钾、结扎线、耳号钳、断齿钳等。

1.2　Delivery tools

The following instruments should be prepared during delivery: record cards for farrowing, surgical scissors, towels, iodine tincture, potassium permanganate, ligation line, forceps for marking ear and breaking teeth, etc.

二、助产方法
2　Midwifery methods

仔猪产出后，立即用清洁毛巾擦去其口、鼻中的黏液，然后再擦干全身。断脐后，进行称重、编号、剪齿、断尾等，并做好记录。

After the piglet is born, wipe the mucus from the mouth, nose and whole body. A lot of work should be done immediately after cutting umbilical cord, such as weighing, numbering, cutting teeth and tail, recording ect..

三、难产救助方法
3　Rescue methods for dystocia

母猪一般不发生难产，如果出现反

Sows generally do not suffer from dystocia and

复阵痛、努责、呼吸及心跳加快等症状即为难产。发现母猪难产时，应马上进行助产，常用"推、拉、注、掏、剖"五字助产技术。

1. 推

如果胎位不正，可采取"推"的办法。即接产者用双手托住母猪后腹部，随母猪努责时向臀部方向用力推，但不可硬压，调整好胎位后再行助产。

2. 拉

看见仔猪的头或腿部时出时进时，只要胎位正，就可以拉出仔猪。

3. 注

母猪分娩过程中如出现产力不足，可肌内注射催产素 10～20 IU，20～30 min 起效。

4. 掏

产不出仔猪时，需要人工掏出。趁母猪努责间歇时，接产员将手指合拢成圆锥形伸入母猪产道，摸到仔猪的适当部位，将仔猪慢慢拉出。如果有两头仔猪同时挤在一起，先将其中一头推回，再抓住另一头随母猪努责拉出，掏出一头仔猪后，如转为顺产，就可不再继续掏。掏完后，用手把 40 万 IU 的青霉素抹入阴道内，以防阴道炎。

5. 剖

采取上述方法后仔猪还生不出来，就应进行剖宫产术。

四、新生仔猪护理

1. 注意观察脐带

脐带断端一般于仔猪出生后 1 周左右干缩脱落。此期应注意观察，勿

the symptoms of dystocia include repeated pains, straining, rapid breathing and heartbeat, etc. If it occurs, the midwifery should assist immediately by using some techniques, which can be summarized as "push, pull, inject, dig and caesarean".

3.1 Push

This method can be used if the fetal position is not correct. Midwives hold the postabdomen of the sow with both hands and push the piglets to the buttocks with the straining of sows to adjust the fetal position.

3.2 Pull

Piglets can be pulled out when the fetal position is right, especially when the head or leg of the fetus coming in and out repeatedly.

3.3 Inject

If the force of labour is insufficient during delivery, oxytocin can be injected intramuscularly for 10-20 IU and it will take effect in 20-30 minutes.

3.4 Dig

If the piglet can not be delievered, midwives need to take them out. Midwives close their fingers into a conical shape and extend into the birth canal during the contracture interval of sows, touch the appropriate parts of the piglet and slowly pull them out. If two piglets are squeezed together, midwives should send one back to the abdomen first, then grab the other one and pull it out with the straining of sows. If the sows turn to normal labour, midwives can stop digging. After digging out, 400 000 IU of penicillin should be wiped into the vagina by hand to prevent vaginitis.

3.5 Caesarean

If the piglet delivery is still impossible after using the aforementioned methods, cesarean section should be performed.

4 Nursing of newborn piglets

4.1 Pay attention to the umbilical cord

It takes about one week after parturition that the broken ends of umbilical cord shrink and abscise.

使仔猪间互相舔吮引起感染发炎，如脐血管或脐尿管关闭不全，应进行结扎处理。

2. 早吃初乳

新生仔猪尽早吃好、吃足初乳，最晚不得超过生后 2h。吃乳前母猪乳头用 0.1％高锰酸钾擦洗消毒，用手挤压各乳头弃去最初挤出的乳汁，然后给仔猪吮吸。

3. 保温

初生仔猪皮薄毛稀，皮下脂肪少，体温调节能力差，对极端温度反应敏感。尤其是在冬季，应密切注意防寒保温，确保产房温度适宜。

4. 假死仔猪急救

个别仔猪产出后不呼吸，但心脏仍在跳动，手指轻压脐带根部可摸到脉搏。急救方法如下：

（1）先除去仔猪口腔、鼻腔内的黏液，然后两手反复伸屈仔猪的两前肢和后肢，直到有呼吸为止。

（2）向假死仔猪鼻内或嘴内用力吹气，促其呼吸。

（3）用左手提起仔猪两后肢，头向下，再用右手拍胸敲背，可救活仔猪。

（4）也可用酒精、碘酊、氨水等涂在仔猪鼻端，刺激其鼻腔黏膜恢复呼吸。

5. 仔猪寄养

为了充分利用生产母猪的所有有效乳头，最大限度地保持吃乳仔猪群的均匀度，需要将一窝中超过母猪乳头数的仔猪进行寄养。寄养前要吃足初乳，一

During this period, piglets should not be allowed to lick each other to prevent infection and inflammation. If the shrink of umbilical vessel and urachus is incomplete, ligation should be performed.

4.2 Feed colostrum as early as possible

The newborn piglets should eat enough colostrum no more than 2 hours after parturition. The nipples of the sow are scrubbed and disinfected with 0.1% potassium permanganate before feeding, and should be squeezed by hand to discard the original milk, before being sucked by piglets.

4.3 Keep warm

Newborn piglets have thin fur, less subcutaneous fat, poor body temperature regulation and are sensitive to extreme temperature. Especially in winter, midwives should pay more attention to ensure that the temperature of parturition room is appropriate.

4.4 Treatment of thanatoid piglets

Some piglets do not breathe after parturition, but their hearts are still beating. A pulse can be felt by pressing finger on the root of the umbilical cord gently. First aid methods are as follows:

(1) First, remove the mucus from the mouth and nasal cavity, then stretch and flex the forelimbs and hind limbs of the piglet repeatedly until it is breathing.

(2) Blow air into the nose or mouth of thanatoid piglets to stimulate them breathing.

(3) Hang the hind limbs of piglets upside down with one hand and gently pat its back and chest with the other hand.

(4) Smear alcohol, iodine tincture and ammonia water on the nasal tip of piglets to stimulate nasal mucosa to restore breathing.

4.5 Foster nursing of piglets

If the litter number of sows exceeds the number of nipples in sows, foster nursing is required. Midwives should feed enough colostrum and smear milk or urine of nurse's sows to the piglets before foster nursing. Foster nursing is generally performed about 2-3

般在出生后 2~3 d 进行。寄养前，在寄养仔猪身上涂抹保姆母猪的乳汁或尿液。

6. 固定乳头

为了减少仔猪在吸乳时的争斗，提高仔猪均匀度，促使母猪乳头充分发育，提高泌乳量，要及时固定乳头。一般母猪的第一对胸部乳头泌乳最多，第2~6 对乳头的泌乳量依次递减，因此，在出生后的第一次吸乳时就要初步固定乳头，让强仔靠后，弱仔靠前。

7. 剪齿

仔猪出生时就有上下各两对锐利的犬齿，如不及时修剪，常因仔猪争抢乳头而把母猪的乳头咬伤。同时，仔猪也因抢占乳头相互咬伤而感染。

8. 定位训练

新生仔猪在生后一至数小时，最晚第二天就会排泄粪尿。在预定排粪地点，放置少许母猪粪便，可诱导仔猪排粪定位。仔猪排粪尿多在吃乳前后，应在此时注意训练定位排粪。

五、产后母猪的护理

产后 2~3 d 内，喂料不宜过多，应用易消化的饲料调成粥状饲喂。喂量逐渐增加，经 5~7 d 后可按哺乳母猪的标准饲喂。母猪分娩后及时清洗消毒其乳房和外阴部，保持产房的清洁卫生。同时应保持产房安静，保证母猪有充分的休息时间。

days after parturition.

4.6 Fixed nipple

In order to reduce the struggle of piglets during sucking, and improve the evenness for each piglet to get fed properly, as well as promote the full development of nipples and increase the lactation volume, it is necessary to fix nipples in time. Generally, the milk yield of the first pair nipples of sows is the most, the milk yield of the second to sixth pairs of nipples decreases successively. Therefore, the nipples should be fixed initially during the first lactation at birth, that arrange the strong piglets behind and the weak piglets ahead.

4.7 Teeth clipping

Piglets are born with two pairs of sharp canine teeth. If the canine teeth are not clipped in time, the piglets often bite the nipples of the sow as they fight for the nipples. At the same time, piglets also bite each other because of fighting for nipples, thus leading to infection with sores.

4.8 Orientation training

Newborn piglets excrete feces and urine from several hours to the next day after birth. A little excrement of sows is placed at the designated place, that can induce the location of excrement of piglets. The best time for the location of defecation training is before and after the piglets sucking milk.

5 Nursing of postpartum sows

The digestible forage should be manufactured into mashy porridge-like food to feed sows within 2-3 days after parturition, and then the feeding quantity increases gradually.

After 5-7 days, the feed can be fed according to the standard of lactating sows. After delivery, breasts and genitals of the sows should be cleaned and disinfected, as well as keeping the delivery room clean. At the same time, the delivery room should be kept quiet, so that the sows can have a good rest.

项目七　繁殖力评价
Project Ⅶ　Evaluation of Fecundity

◆ 项目导学

繁殖力是评定种用畜禽生产力的主要指标，合理评价畜禽的繁殖力，能客观了解畜禽的繁殖状况，为制定繁殖管理措施提供依据。

◆ Project Guidance

Fecundity is the main indicator to evaluate the productivity of breeding animals. Reasonable evaluation of fertility can objectively understand the reproductive status and provide a basis for formulating reproductive management measures.

◉ 学习目标

>>> 知识目标

- 理解繁殖力评定指标的含义及统计方法。
- 熟悉各种畜禽的正常繁殖力。

>>> 技能目标

- 能根据养殖场的繁殖资料，统计分析各繁殖力指标，并进行综合评价。
- 结合生产实际，拟定养殖场的繁殖管理措施。

◉ Learning Objectives

>>> Knowledge Objectives

- To understand the meaning and statistical methods of fertility assessment indicators.
- Be familiar with the normal fecundity levels of various livestock and poultry.

>>> Skill Objectives

- According to the breeding data of the farm, analyze various reproductive indicators and evaluate comprehensively.
- Combining with the actual production, formulate the breeding management measures of the farm.

▲ 相关知识

一、繁殖力

繁殖力是指畜禽维持正常生殖机能、繁衍后代的能力。对于种用畜禽来说，繁殖力即是生产力。公畜（禽）的繁殖力主要体现在性成熟早晚、性欲强

▲ Relevant Knowledge

1　Fecundity

Fecundity refers to the ability of livestock and poultry to maintain normal reproductive function and reproduce offspring. For breeding animals, fecundity is productivity. Fecundity of male animals is mainly re-

弱、交配能力及精液品质等，母畜（禽）的繁殖力主要体现在性成熟早晚、发情排卵情况以及配种妊娠、分娩和哺乳等生殖活动。

二、繁殖力的评定指标

（一）家畜繁殖力评定指标

1. 受配率

本年度内参加配种的母畜数占适繁母畜数（不包括因妊娠、哺乳及各种卵巢疾病等原因造成空怀的母畜）的百分率。

2. 受胎率

受胎率是指妊娠母畜数占参加配种母畜数的百分率。在受胎率统计中又分为总受胎率、情期受胎率、第一情期受胎率和不返情率。

总受胎率指本年度妊娠母畜数占本年度内参加配种母畜数的百分率。情期受胎率指妊娠母畜数占配种情期数的百分率。第一情期受胎率指第一情期配种的妊娠母畜数占第一情期配种母畜数的百分率。不返情率指配种后一定时间内（如 30d、60d、90d 等）未出现发情的母畜数占参加配种母畜数的百分率。

3. 分娩率

分娩率是指本年度内分娩母畜数占妊娠母畜数的百分率。其大小反映母畜妊娠质量的高低和保胎效果。

4. 产仔率

产仔率是指母畜的产仔（包括死胎）数占分娩母畜数的百分率。单胎动物如牛、马、驴等，产仔率一般不会超过 100%，生产上多使用分娩率。多胎

flected in sexual maturation, sexual desire, mating ability and semen quality. Fecundity of female animals is mainly reflected in sexual maturation, estrus and ovulation, pregnancy, delivery and lactation.

2 Indicators of evaluating fecundity

2.1 Indicators for evaluating livestock fecundity

2.1.1 Breeding rate

Breeding rate refers to the number of bred females accounted for brood matron in one year (excluding unpregnant females that are caused by pregnancy, lactation and various ovarian diseases).

2.1.2 Conception rate

Conception rate refers to the number of pregnant females accounted for the bred females. In the statistics of conception rate, it can be divided into total conception rate, cycle conception rate, first cycle conception rate and non-return rate.

Total conception rate refers to the number of pregnant females accounted for bred females in one year. Cycle conception rate refers to the number of pregnant females accounted for total cycles. First-cycle conception rate refers to the number of pregnant females in the first cycle accounted for the females bred in the first cycle. Non-return rate refers to the number of females who do not estrus within a certain period of time after breeding (30d, 60d, 90d, etc.) accounted for bred females.

2.1.3 Delivery rate

It refers to the number of delivered females accounted for pregnant females in the current year. It reflects the quality of pregnancy and the effect of fetal protection.

2.1.4 Farrowing rate

It refers to the number of the newborns (including stillbirths) accounted for the delivered females. Single-born animals (such as cattle, horses, donkeys, etc.) generally do not exceed 100% of the litter size. Delivery rate is often used in production rather than farrowing

项目七 繁殖力评价
Project Ⅶ Evaluation of Fecundity

动物如猪、山羊、兔等，产仔率均会超过 100%，生产上多使用产仔率。

5. 仔畜成活率

仔畜成活率指本年度内断奶成活的仔畜数占本年度产出活仔畜数的百分率。其大小反映仔畜的培育情况。

（二）家禽繁殖力指标的统计

1. 产蛋量

产蛋量是指某群家禽一年内平均产蛋的枚数。

2. 受精率

受精率是指种蛋孵化后，经第一次照蛋确定的受精蛋数占入孵蛋数的百分率。

3. 孵化率

孵化率可分为受精蛋的孵化率和入孵蛋的孵化率两种，分别指出雏数占受精蛋数和入孵蛋数的百分率。

4. 育雏率

育雏率是指育雏期末成活雏禽数占入舍雏禽数的百分率。

三、畜禽的正常繁殖力

1. 牛的正常繁殖力

由于不同地区的饲养管理条件、繁殖管理水平和环境气候差异等原因，牛群的繁殖力水平也有很大差异。我国奶牛的繁殖力低于发达国家水平，一般成年母牛的情期受胎率为 40%～60%，全年总受胎率为 75%～95%，分娩率为 93%～97%，年繁殖率为 70%～90%，产犊间隔为 12～14 个月，双胎率为 3%～4%，繁殖年限一般为 8～10 年。

rate. Multiparous animals(such as pigs, goats, rabbits, etc.), will exceed 100% of the farrowing rate. Farrowing rate is usually used in production.

2.1.5 Survival rate

It refers to the number of weaning newborns accounted for newborns in a year. It reflects the cultivation of the litter.

2.2 Indicators for evaluating poultry fecundity

2.2.1 Egg production

Egg production refers to the average number of eggs laid by a flock of poultry in one year.

2.2.2 Fertilization rate

After incubating, the number of fertilized eggs determined by the first candled accounted for the number of incubated eggs.

2.2.3 Hatchability

It can be divided into two types: the hatchability of fertilized eggs and the hatchability of incubated eggs. They refer to the number of chicks accounted for the number of fertilized eggs and the number of incubated eggs, respectively.

2.2.4 Brooding rate

Brooding rate refers to the number of young birds alive at the end of the brooding time accounted for the total number of young birds in the brood-grow house.

3 Normal fecundity of livestock and poultry

3.1 Normal fecundity of cattle

In different regions, the fecundity level of cattle is also very different due to the differences of feeding management, reproductive management, environment and climate. The fecundity of cows in China is lower than that of developed countries. The cycle conception rate is 40%-60%, the total conception rate is 75%-95%, the delivery rate is 93%-97%, the annual reproduction rate is 70%-90%, the calving interval is 12-14 months, the twin rate is 3%-4%, and the breeding period is 8-10 years.

2. 羊的正常繁殖力

在自然交配的情况下，种公羊一般能为30~50只母羊配种，采用人工授精可将配种能力提高数千倍。绵羊大多一年一胎或两年三胎，产单羔。在饲养条件较好的地区，可产双羔、三羔或者更多，其中湖羊繁殖率最强，其次为小尾寒羊，平均每胎产羔2只以上，最多可达7~8只，2年可产3胎或年产2胎。山羊一般每年产羔1胎，每胎产羔1~3只，个别可产羔4~5只。羊的受胎率均在90%以上，情期受胎率约为70%，繁殖年限为8~10年。

3. 猪的正常繁殖力

猪的繁殖力很高，中国猪种一般产仔10~12头，太湖猪平均14~17头，个别可以产25头以上，年平均产仔窝数为1.8~2.2窝。母猪正常情期受胎率为75%~80%，总受胎率为85%~95%，繁殖年限为8~10年。正常断奶情况下，每年可产2.2~2.5窝。

4. 家禽的正常繁殖力

蛋鸡的产蛋量一般为250~300枚，肉鸡150~180枚，蛋鸭200~250枚，肉鸭100~150枚，鹅30~90枚。蛋的受精率一般在90%以上，受精蛋孵化率在80%以上，入孵蛋孵化率在65%以上，育雏率一般达到80%~90%。

3.2 Normal fecundity of sheep and goat

In natural mating, a breeding ram can breed 30-50 ewes, and artificial insemination can increase the breeding ability thousands of times. Most sheep have one birth in one year or three births in two years and give a single lamb per litter. In some areas with better feeding conditions, twin lambs, three lambs or more can be gave per litter. Among which the hu sheep has the strongest fecundity, followed by small tail han sheep, with an average of more than 2 lambs per litter, up to 7-8 lambs per litter at most. Small-tail han sheep have three births in two years or two births in one year. Goats usually have one birth in one year, 1-3 lambs per litter, and up to 4-5 lambs per litter sometimes. The conception rate of sheep is above 90%, the cycle conception rate is about 70%, and the breeding period is 8-10 years.

3.3 Normal fecundity of pig

The fecundity of pigs is very high. Chinese pig breeds generally have 10-12 piglets per litter, taihu pigs have an average of 14-17 piglets, and some individuals can have more than 25 piglets per litter. The average litters per sow per year (LSY) is 1.8-2.2. The cycle conception rate of sows in normal conditions is 75%-80%, the total conception rate is 85%-95%, and the breeding period is 8-10 years. Under normal weaning conditions, the LSY is approximately 2.2-2.5.

3.4 Normal fecundity of poultry

Layers generally produce 250-300 eggs per year, 150-180 in broilers, 200-250 in layer ducks, 100-150 in broiler ducks and 30-90 in geese. The fertilization rate is generally above 90%, the hatchability of fertilized eggs and the hatchability of incubated eggs are above 80% and 65%, respectively. And the brooding rate is generally 80%-90%.

任务 繁殖力评价
Task　Evaluation of Fecundity

任务描述
Task Description

在畜禽的繁殖工作中，要熟悉养殖场的繁殖操作流程，认真做好繁殖资料的记录和管理工作，定期进行统计分析，为解决生产实际问题和制订繁殖计划提供第一手资料。

In the breeding work of animals, it is necessary to familiarize with the breeding process of the farm, record and manage the breeding data carefully, and conduct statistical analysis regularly, so that provide first-hand information for solving practical production problems and formulating breeding plans.

▶案例1：表7-1是某奶牛场的繁殖数据，请计算受配率、总受胎率、情期受胎率、第一情期受胎率、30d不返情率、分娩率、产犊率及年繁殖率。

▶Case 1: Table 7-1 is a data of a dairy farm, please calculate the following indicators: breeding rate, total conception rate, cycle conception rate, first-cycle conception rate, non-return rate, breeding index, delivery rate, calving rate, annual reproductive rate.

表7-1　某奶牛场的繁殖数据
Table 7-1　Data of a dairy farm

项　目 Items	数　量 Number	备　注 Remarks
存栏母牛 Total cows	2 025	
适繁母牛 Brood cows	1 612	
配种母牛 Bred cows	1 571	正常发情 Normal estrus
配种后30d发情母牛 Cows estrus within 30 days after breeding	477	
配种后80d妊娠母牛 Pregnant cows (80 days after breeding)	1 495	978头牛第1次发情配种受孕 978 cows pregnant after the first service 365头牛第2次发情配种受孕 365 cows pregnant after the second service 152头牛第3次发情配种受孕 152 cows pregnant after the third service
分娩母牛 Delivered cows	1 432	
产犊数 Calves	1 466	

▶ 案例2：某猪场，在2018年内有繁殖母猪500头，共繁殖了1 180窝仔猪，出生的仔猪头数为13 580头，其中有98头死胎和木乃伊胎儿，到28日龄断奶时存活了12 996头。请计算该猪场在2018年度内的平均产仔窝数、平均窝产仔数、产活仔数及仔猪成活率。

▶ 案例3：某种鸡场繁殖数据统计见表7-2，试评价该鸡场本产蛋期的繁殖力。

▶ Case 2: In a pig farm in 2018, 500 brood sows were bred with 1 180 litters. There were 13 580 newborn piglets, of which 98 were stillbirths and mummified fetus, and the number of survivors at 28-day-old weaning was 12 996. Please calculate average litters per sow per year (LSY), litter size, number of piglets alive, survival rate.

▶ Case 3: Table 7-2 is a data of a layer farm, please calculate the fecundity indicators.

表 7-2 某种鸡场的繁殖数据
Table 7-2 Data of a layer farm

项　目 Items	数量（万） Number (ten thousands)	备　注 Remarks
产蛋种鸡 Total layers	0.5	
产蛋数 Total eggs	239	
入孵蛋 Hatchable eggs	230	合格种蛋 Qualified eggs
未受精蛋 Unfertilized eggs	3.586	
出雏数 Chicks	203.573 5	
健雏鸡 Young chicks alive	203.15	育雏结束 Alive after brooding period

任务实施（Task Implementation）

▶ 案例1（Case 1）

1. 受配率（Breeding rate）

$$受配率（breeding\ rate）= \frac{配种母牛数（No.\ of\ bred\ females）}{适繁母牛数（No.\ of\ brood\ matron）} \times 100\%$$

$$= \frac{1\ 571}{1\ 612} \times 100\% = 97.46\%$$

2. 受胎率 (Total conception rate)

总受胎率 (total conception rate) $= \dfrac{\text{本年度妊娠母牛数 (No. of pregnant females)}}{\text{本年度内参加配种的母牛数 (No. of bred females)}} \times 100\%$

$= \dfrac{1\ 495}{1\ 571} \times 100\% = 95.16\%$

3. 情期受胎率 (Cycle conception rate)

情期受胎率 (cycle conception rate) $= \dfrac{\text{妊娠母牛数 (No. of pregnant females)}}{\text{配种情期数 (No. of total cycles)}} \times 100\%$

$= \dfrac{1\ 495}{1\ 571 + (1\ 571 - 978) + (1\ 571 - 978 - 365)} \times 100\%$

$= 62.50\%$

4. 第一情期受胎率 (First cycle conception rate)

第一情期受胎率 (first cycle conception rate) $=$

$\dfrac{\text{第一情期配种妊娠母牛数 (No. of pregnant females first service)}}{\text{第一情期配种母牛数 (No. of bred females first service)}} \times 100\%$

$= \dfrac{978}{1\ 571} \times 100\% = 62.25\%$

5. 不返情率 (Non-return rate)

x d 不返情率 (non-return rate after x days)

$= \dfrac{\text{配种后 } x \text{ d 未返情母牛数 (No. of non-return females after } x \text{ days)}}{\text{配种母牛数 (No. of bred females)}} \times 100\%$

30d 不返情率 (non-return rate after 30 days) $= \dfrac{1\ 571 - 477}{1\ 571} \times 100\% = 69.64\%$

6. 分娩率 (Delivery rate)

分娩率 (delivery rate) $= \dfrac{\text{分娩母牛数 (No. of delivered females)}}{\text{妊娠母牛数 (No. of pregnant females)}} \times 100\%$

$= \dfrac{1\ 432}{1\ 495} \times 100\% = 95.79\%$

7. 产犊率 (Calving rate)

产犊率 (calving rate) $= \dfrac{\text{产出犊牛数 (No. of calves born)}}{\text{分娩母牛数 (No. of delivered females)}} \times 100\%$

$= \dfrac{1\ 466}{1\ 432} \times 100\% = 102.37\%$

8. 年繁殖率 (Annual reproductive rate)

年繁殖率 (annual reproductive rate) $= \dfrac{\text{本年度实繁母牛数 (No. of delivered females)}}{\text{本年度适繁母牛数 (No. of brood females)}} \times 100\%$

$= \dfrac{1\ 432}{1\ 612} \times 100\% = 88.83\%$

▶ **案例 2 (Case 2)**

1. 产仔窝数 (LSY)

年产仔窝数(窝) [LSY(Litters)] $= \dfrac{\text{年度内分娩的窝数 (No. of annual litters)}}{\text{年度内繁殖母猪数 (annual No. of brood sows)}} \times 100\%$

$$=\frac{1\ 180}{500}\times 100\% = 2.36\ \text{(litters)}$$

2. 窝产仔数 (Litter size)

$$\text{窝产仔数（头）[litter size(number of piglets)]} = \frac{\text{年度内的产仔总数(annual litter size)}}{\text{年度内的产仔窝数(annual No. of LSY)}}\times 100\%$$

$$=\frac{13\ 580}{1\ 180}\times 100\% = 11.51$$

3. 产活仔数 (Number of piglets alive)

产活仔数（头）(number of piglets alive) = 出生的仔猪数（No. of piglets born）− 死胎和木乃伊胎儿数量（No. of stillbirth and mummified fetus）= 13 580 − 98 = 13 482

4. 仔猪成活率 (Survival rate)

$$\text{仔猪成活率（survival rate）} = \frac{\text{断奶成活仔猪数（No. of weaning newborns）}}{\text{出生活仔猪数（No. of newborns alive）}}\times 100\%$$

$$=\frac{12\ 996}{13\ 482}\times 100\% = 96.39\%$$

▶ 案例 3（Case 3）

1. 种蛋合格率 (Hatchable egg rate)

$$\text{种蛋合格率（hatchable egg rate）} = \frac{\text{合格种蛋数（No. of hatchable eggs）}}{\text{产蛋总数（total eggs）}}\times 100\%$$

$$=\frac{2\ 300\ 000}{2\ 390\ 000}\times 100\% = 96.23\%$$

2. 受精率 (Fertilization rate)

$$\text{受精率（fertilization rate）} = \frac{\text{受精蛋数（No. of fertilized eggs）}}{\text{入孵蛋数（No. of incubated eggs）}}\times 100\%$$

$$=\frac{2\ 300\ 000 - 35\ 860}{2\ 300\ 000}\times 100\% = 98.44\%$$

3. 孵化率 (Hatchability)

$$\text{受精蛋孵化率（hatchability of fertilized eggs）} = \frac{\text{出雏数（No. of young chicks born）}}{\text{受精蛋数（No. of fertilized eggs）}}\times 100\%$$

$$=\frac{2\ 035\ 735}{2\ 300\ 000 - 35\ 860}\times 100\% = 89.91\%$$

$$\text{入孵蛋孵化率（hatchability of incubated eggs）} = \frac{\text{出雏数（No. of young chicks born）}}{\text{入孵蛋数（No. of incubated eggs）}}\times 100\%$$

$$=\frac{2\ 035\ 735}{2\ 300\ 000}\times 100\% = 88.51\%$$

4. 育雏率 (Brooding rate)

$$\text{育雏率（brooding rate）} = \frac{\text{育雏期末成活雏禽数（No. of young chicks alive）}}{\text{入舍雏禽数（total young chicks in brood）}}\times 100\%$$

$$=\frac{2\ 031\ 500}{2\ 035\ 735}\times 100\% = 99.79\%$$

REFERENCES 参考文献

陈炫华，黄其敏，2014. 提高牛繁殖力的饲养管理措施[J]. 中国畜牧兽医文摘，1：37.
耿明杰，2006. 畜禽繁殖与改良[M]. 北京：中国农业出版社.
耿明杰，常明雪，2013. 动物繁殖技术[M]. 北京：中国农业出版社.
李波，2014. 奶牛繁殖管理[J]. 四川畜牧兽医，9：42-43.
李长彬，2013. 牛冷冻精液的制作程序[J]. 养殖技术顾问，4：58.
李凤玲，2011. 动物繁殖技术[M]. 北京：北京师范大学出版社.
刘海良，金穗华，张晓霞，2012. 牛精液冷冻技术与质量控制[J]. 畜牧与兽医，44（4）：88-91.
刘士文，2012. 奶牛精液的冷冻保存[J]. 养殖技术顾问，9：47.
宋亚攀，孙丽萍，郭爱珍，等，2014. 涂蜡笔方法在母牛发情鉴定中的应用[J]. 中国奶牛，15：60-62.
王锋，2006. 动物繁殖学实验教程[M]. 北京：中国农业大学出版社.
王锋，2012. 动物繁殖学[M]. 北京：中国农业大学出版社.
王建，2013. 鸡人工授精技术操作要点[J]. 黑龙江动物繁殖，21（5）：28-30.
徐相亭，秦豪荣，等，2008. 动物繁殖[M]. 北京：中国农业大学出版社.
徐占晨，2009. 浅谈种公猪精液的稀释、保存和运输[J]. 黑龙江动物繁殖，17（2）：46-47.
杨利国，2003. 动物繁殖学[M]. 北京：中国农业出版社.
张响英，2018. 动物繁殖技术[M]. 北京：中国农业出版社.
钟孟淮，2015. 动物繁殖与改良[M]. 北京：中国农业出版社.
朱鸿昌，毛蕾，王绍伟，2009. 解析牛冷冻精液新国标[J]. 河南畜牧兽医，30（4）：23.
Hafez B, Hafez E S E, 2000. Reproduction in Farm Animals [M]. Philadelphia：Lippincott Williams & Wilkins co.
Richard M Hopper, 2015. Bovine Reproduction [M]. Hoboken：John Wiley & Sons, Inc.
Perry T Cupps, 1991. Reproduction in Domestic Animals [M]. New York：Academic Press, Inc.
Ian Gordon, 2005. Reproductive Technologies in Farm Animals [M]. Oxfordshire：CABI Publishing.
Richard M, 2014. Bourdon Understanding Animal Breeding [M]. London：Pearson Education Limited.
Blackwelder, Richard Eliot, 2018. The Diversity of Animal Reproduction [M]. Boca Raton：CRC Press, Inc.

图书在版编目（CIP）数据

动物繁殖/张响英，杨晓志主编．—北京：中国农业出版社，2021.11
高等职业教育农业农村部"十三五"规划教材 "十三五"江苏省高等学校重点教材
ISBN 978-7-109-28270-4

Ⅰ.①动… Ⅱ.①张…②杨… Ⅲ.①动物—繁殖—高等职业教育—教材 Ⅳ.①S814

中国版本图书馆 CIP 数据核字（2021）第 093219 号

动物繁殖
Animal Reproduction

中国农业出版社出版
地址：北京市朝阳区麦子店街 18 号楼
邮编：100125
责任编辑：李 萍　文字编辑：陈睿赜
版式设计：王 晨　责任校对：吴丽婷
印刷：三河市国英印务有限公司
版次：2021 年 11 月第 1 版
印次：2021 年 11 月河北第 1 次印刷
发行：新华书店北京发行所
开本：787mm×1092mm 1/16
印张：12.25
字数：290 千字
定价：38.00 元

版权所有·侵权必究
凡购买本社图书，如有印装质量问题，我社负责调换。
服务电话：010-59195115　010-59194918